Does This Line Ever Move?

Everyday Applications of Operations Research

Kenneth R. Chelst • **Thomas G. Edwards**

Department of Industrial and
Manufacturing Engineering
Wayne State University

College of Education
Wayne State University

Key Curriculum Press
Innovators in Mathematics Education

Editor: Ladie Malek
Editorial Assistant: Kristin Burke
Consulting Editor: Josephine Noah
Teacher Reviewer: Deborah Davies
Accuracy Checker: Dudley Brooks
Production Director: Diana Jean Ray
Production Editor: Angela Chen
Copyeditor: Rebecca Pepper
Production Coordinator, Compositor: Ann Rothenbuhler
Text Designer: Marilyn Perry
Art Editor: Jason Luz
Art and Design Coordinator: Kavitha Becker
Cover and Unit Opener Designer: Todd Bushman
Photo Credits: Cover—Main image (people in line): Picturequest; Girl/Food Service: Don
    Tremain/Picturequest; Man/Airport Counter: Creatas; Rolls of Paper: Ken Davies/Masterfile;
    Freeway Scene/Todd Bushman; Interior—**1**: London Aerial Photo Library/Corbis; **3**: Durand
    Patrick/Corbis Sygma; **13**: Corbis; **30**: Corbis; **43**: Annie Griffiths Belt/Corbis; **55**: Julius/Corbis;
    65: Erik Freeland/Corbis Saba; **75**: Anthony Redpath/Corbis; **89**: Chuck Savage/Corbis;
    107: Electronic Arts Inc.; **127**: Chuck Savage/Corbis
Prepress, Printer: Versa Press, Inc.

Executive Editor: Casey FitzSimons
Publisher: Steven Rasmussen

Key Curriculum Press
1150 65th Street
Emeryville, CA 94608
editorial@keypress.com
www.keypress.com

Printed in the United States of America
10 9 8 7 6 5 4 3 2 1   08 07 06 05 04
ISBN: 1-55953-673-X

Contents

    Finding the shortest route
        *Students become familiar with the basic elements of a network, develop
        mathematical reasoning, and apply a shortest-path algorithm.*

    A traveling salesman problem
        *Students compare the efficiency and effectiveness of brute force and
        heuristic algorithms.*

    Optimizing the product mix to maximize production
        *Students identify the constraints and the objective function in a
        manufacturing context, and discover the reasoning behind the
        corner principle.*

    Finding the optimal cutting pattern
        *Students learn about integer programming and find the lattice point
        within the feasible region that represents the optimal solution.*

Introduction

We began this project in 1996 with a very simple goal: to use the rich real-world contexts of operations research to better motivate high school students to learn mathematics. As the project developed, we realized that those same real-world contexts can also improve students' understanding of mathematical ideas. When students are familiar with the context, they can more readily conceptualize the content.

Operations research has numerous applications in everyday situations. Students will be familiar with many of the contexts, and this familiarity supports the development of the related mathematical concepts.

The materials in *Does This Line Ever Move?* were developed to align with the National Council of Teachers of Mathematics Standards—first the *Curriculum and Evaluation Standards for School Mathematics* (1989), and later the *Principles and Standards for School Mathematics* (2000). In particular, the activities in this book

- carefully develop one or two big ideas, which are then connected and extended to other mathematics or real-world contexts.

- focus on the usefulness of mathematics as it is applied to real-world problem-solving situations.

- involve worthwhile mathematical tasks that pique students' curiosity and draw them into the lessons. This is especially true because the tasks have been connected to students' life experience, wherever possible.

- support learning with understanding, because students are actively engaged in mathematical tasks and experiences designed to deepen and connect their knowledge of mathematics.

- use technology where appropriate, reflecting the ubiquitous use of technology in the 21st-century workplace.

WHAT IS OPERATIONS RESEARCH?

Operations research (OR) is a scientific approach to analyzing problems and making decisions. OR professionals seek to understand the structure of complex situations, and to use this understanding to predict system behavior and improve system performance. Much of this work is done using analytical and numerical techniques to develop and manipulate mathematical and computer models of organizational systems that involve people, machines, and procedures.

The field of operations research has its roots in the years just prior to World War II, when the British were preparing for the anticipated air war. In 1938, scientists set up experiments to explore how the information provided by the new technology of radar should be used to direct deployment and use of fighter planes. Until this time, the word "experiment" referred to a scientist carrying out a controlled experiment in a laboratory. In contrast, the multi-disciplinary team of scientists working on this radar–fighter plane project studied the actual operating conditions of these new devices and designed experiments in the field of operations. The team's goal was understand the operations of a complete system involving equipment, people, and environmental conditions (such as weather and daylight), and then improve upon it. The new term "operations research" was born.

In the 1950s, operations research evolved into a profession with the formation of national societies and the establishment of journals and academic departments in universities. The use of operations research expanded beyond the military to include both private companies and other governmental organizations. The petrochemical industry was one of the first to broadly embrace operations research in order to improve the performance of refineries, develop natural resources, and plan strategy. Today, operations research plays important roles in a variety of industries such as pharmaceuticals, airlines, logistics services, financial services, manufacturing, and all levels of government. As the field evolved, the focus moved away from interdisciplinary teams to the development of mathematical models that can be used to model, improve, and even optimize real-world systems.

Operations researchers use data and analytical tools. However, sometimes it is also instructive to learn how a company in a similar situation solved a business problem. An operations research solution to a business problem often involves a combination of OR tools and other changes and innovations. For this reason, operations researchers frequently make use of **case studies**—detailed analyses of business challenges and how they were solved. Case studies also offer business decision makers real-world examples analogous to their own, and illustrate the potential value of applying OR in their context. Most of the case studies in this book were originally published in the operations research journal *Interfaces*. We hope that they will illustrate for your students the uses of operations research techniques in everyday business situations.

Operations researchers have interests that overlap with those of professionals in many other disciplines. Research into algorithms often parallels the work of computer scientists. Work in the area

of quality and reliability applies to a number of engineering disciplines. Forecasting models, an important element of OR-based decision support systems, are also of interest to statisticians and economists. Lastly, OR specialists in applied mathematical programming or game theory may work with colleagues in economics. Not surprisingly, many Nobel Prize winners in economics have had strong links with the operations research community. These include Kenneth J. Arrow (1972), Wassily Leontief (1973), Leonid Vitaljevich Kantorovich and Tjalling Charles Koopmans (1975), Herbert A. Simon, (1978), George Stigler (1982), Maurice Allais (1988), Harry M. Markowitz (1990), and John F. Nash (1994).

For more information about the field of operations research, explore our website, at **www.hsor.org**; the INFORMS (Institute for Operations Research and the Management Sciences) website, **www.informs.org**; the British Operational Research Society website **www.theorsociety.org.uk**; and the *Encyclopedia of Operations Research and Management Science, 2nd edition*, S. I. Gass and C. M. Harris, editors.

Kenneth R. Chelst
Thomas G. Edwards

References

Beasley, J. E., http://mscmga.ms.ic.ac.uk/jeb/or/intro.html.

Brothers, L. A. 1954. Operations Analysis in the United States Air Force. *Operations Research* 2:1–16.

Eppen, G. D., F. J. Gould, C. P. Schmidt, J. H. Moore, and L. R. Weatherford, *Introductory Management Science, 5th edition* (Upper Saddle River, NJ: Prentice Hall, 1998), 2–24.

Kirby, M. W. and R. Capey. 1997. The Air Defense of Great Britain, 1920–1940: An Operational Research Perspective. *JORS* 48:661–667.

Kirby, M. W. 2001. History of Early British OR. *Encyclopedia of Operations Research and Management Science, 2nd edition*, Gass, S. I. and Harris, C. M., editors.

McCloskey, J. F. 1989. US Operations Research in World War II. *Operations Research* 35:910–925.

ACKNOWLEDGMENTS

We would like to thank the Department of Industrial and Manufacturing Engineering at Wayne State University and the Department of Operations Research at George Mason University for the use of their computer facilities. We are also grateful to a number of colleagues from the field of operations research for their helpful critique and insights, including Matthew Rosenshine, Donald Gross, Jack Pettit, Frank Trippi, Marilyn Maddox, R. Jean Ruth, Karla Hoffman, Dave Goldsman, and the late Carl Harris.

Finally, we are greatly indebted to the teacher writing teams that were instrumental in transforming OR vignettes into classroom-ready materials. The teacher writers who contributed so much to this project are

Lisa Breidenbach	Rachel Carson Middle School, Herndon, Virginia
Ellen Chien	Langley High School, McLean, Virginia
Rhonda Cooke	Denby High School, Detroit, Michigan
Thomas Evasic	Farmington High School, Farmington, Michigan
Deborah Ferry	Oakland County Intermediate School District, Pontiac, Michigan
Ilana Hand	Langley High School, McLean, Virginia
Christine Langley	Detroit Country Day School, Birmingham, Michigan
Marlene Lawson	Governor's School for Government and International Studies, Richmond, Virginia
David Menczer	Southwestern High School, Detroit, Michigan
Hazel Orth	Langley High School, McLean, Virginia
Helen Snyder	NCTM, Reston, Virginia

—K. R. C. and T. G. E.

HOW TO USE THESE ACTIVITIES

The overall objective of these lessons is to motivate students to learn the mathematics they are studying in class. We hope this will occur when students see mathematics applied to everyday life. Aspects of the real-world context have been used in the lesson activities to support student understanding of these concepts. These lessons can be integrated into any first-year or second-year algebra course (or the integrated courses in which students work with literal equations, graphing, or rational equations). The specific mathematics content objectives are detailed in this chart.

Topics	Activity Number									
	1	2	3	4	5	6	7	8	9	10
Writing equations and inequalities			✓	✓	✓	✓	✓			
Evaluating expressions			✓	✓	✓	✓	✓	✓	✓	✓
Domain/range of a function			✓	✓	✓	✓	✓			✓
Asymptotic behavior							✓			
Systems of inequalities			✓	✓	✓	✓				
Finding and interpreting solutions	✓	✓	✓	✓	✓	✓	✓	✓	✓	✓
Data analysis								✓	✓	✓
Probability							✓	✓	✓	
Graph Theory	✓	✓								
Difficulty Level	1	1	2	2	2	3	3	2	2	2

Difficulty levels:
1: No prerequisites
2: Some basic algebra needed
3: Some advanced algebra needed

Each activity is followed by **Teacher Notes** and **Solutions.** For ease of use, **transparency and worksheet masters** are also provided for most activities.

This book contains many **case study summaries.** They accompany each unit opener, and also appear in brief, in the feature **Everyday Applications of Operations Research** at the end of each student activity. These provide a wealth of information about the field of operations research, and on how quantitative techniques are used in the business world. Many more case studies, as well as more projects, extensions, and notes, are available on our **website,** at [**www.hsor.org**].

Routing Through Networks

BACKGROUND

The field of graph theory dates back more than 250 years to the Swiss mathematician Leonhard Euler (1707–1783). Euler used theoretical analysis to prove that it was impossible to cross each of the seven bridges in the town of Königsberg exactly once and return to the starting point.

Operations researchers use graphs to represent and solve practical routing problems. The advancement of computers in the 1950s and 1960s offered the opportunity to solve an array of large-scale routing problems in a reasonable time. A graph was no longer just a collection of **nodes,** or locations, connected by arcs, but also included numeric values to represent either distance or time between nodes. Operations researchers developed **algorithms,** or step-by-step procedures, that utilized the new technology to solve complex routing problems on large networks.

Dutch-American computer scientist Edsger Wybe Dijkstra (1930–2002) developed one of the first efficient algorithms for finding the shortest or fastest route between two specific locations. Today, shortest-path algorithms (such as Dijkstra's) are key building blocks of more sophisticated algorithms used to design and manage complex networks, such as those encountered in the telecommunications industry.

Two common kinds of vehicle routing problems are: the traveling salesman problem, first studied in a general form by 19th-century Irish mathematician William Rowan Hamilton (1805–1865), and the Chinese postman problem, developed by Chinese mathematician Mei-Ko Kwan in 1962 for rural postal routes in China. In the traveling salesman problem, an individual or vehicle travels along the shortest route from node to node, visits every node in the network once, and then returns to home base. In the Chinese postman problem, the traveler must traverse every arc—that is, road link—in the network once. The Chinese postman algorithm, an extension of Euler's Königsberg bridge problem, is relevant to a number of other applications, such as garbage collection, street sweeping, meter reading, and inspecting roads for drivability.

References

Bodin, Lawrence D. 1990. Twenty years of routing and scheduling. *Operations Research* 38:571–579.

A routing problem might involve finding the fastest route to your destination or finding a route that covers every block within a designated area with minimal backtracking.

CASE STUDY: ROUTING DELIVERY AND SERVICE VEHICLES AT SEARS

Department stores sell large furniture and appliances that must be delivered to customers. The salesperson determines the day and estimated time of delivery based on customer preference and the delivery schedule in the area where the customer is located. Some stores also offer installation and repair services on their appliances, which need similar scheduling.

Sears, Roebuck and Company uses a vehicle routing and scheduling system coupled with a geographic information system (GIS) to run its delivery and home service fleets more efficiently. In the late 1990s, operations research consultants helped Sears set up transportation and logistics services as well as product services.

Sears' logistics services department manages a U.S. fleet of over 1000 delivery vehicles to bring the products they sell to the customer's location. Each day, Sears' logistics services employees create the delivery routes for the following day's deliveries. They take into account the types and quantities of merchandise, available vehicles, and customer time windows. The routing managers attempt to maximize the accuracy of the delivery time windows, minimize operational costs, and give drivers consistent routes.

Sears' product services department operates a U.S. fleet of 12,500 service vehicles driven by service technicians who repair and install appliances. Each day, the regional product services employees build service routes based on customer requests for the following day, taking into account the availability of technicians, their skills, and their work schedules. However, these routes may be revised to accommodate emergencies or changes in technicians' work schedules. Operating 13,500 vehicles a day requires some planning. Once Sears began using these new systems, its customer satisfaction rates rose above 80% for both delivery and service, and it was able to save $42 million annually.

References

Weigel, D., and B. Cao. 1999. Applying GIS and OR techniques to solve Sears' technician-dispatching and home-delivery problems. *Interfaces* 29(1):112–130.

➤ For more case studies and background, go to (**www.hsor.org**).

ACTIVITY 1

Service Woes at Speedy Delivery

We can model any travel and delivery problem by using a network representation. A **network** consists of a set of points, called nodes, which are connected by arcs. Here is an example of a simple network.

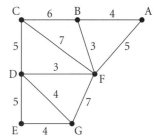

Global Positioning Systems (GPS) are installed in some cars to help the driver find the shortest route.

In practical applications, the nodes often represent geographic points, such as cities, intersections, railroad stops, pipeline connections, or individual locations. The arcs represent links between nodes, often roads between cities. The arcs can be undirected (two way) or directed (one way). Sometimes the arcs are labeled with numeric values representing distance, travel time, or cost. A basic problem involving networks is to find the shortest path (a sequence of nodes and arcs) between two given nodes.

Speedy Delivery has a delivery area represented by this network. The numeric labels indicate the average driving times in minutes between two pickup locations. For example, on average it should take a driver four minutes to drive from location A to location B.

1. On average, how long should it take the driver to travel from A to B to C (path ABC)?

Node A represents the company headquarters, and node E is the location of the company's largest customer.

2. Lorraine Bosco, the dispatcher for Speedy Delivery, wants to find the quickest route from A to E. What route from A to E do you think requires the least time? Compare answers within your group.

As you can see, there are a variety of paths and times from A to E. There is a systematic method, called an algorithm, for finding the fastest route. Dijkstra's

Algorithm is a shortest-path algorithm. You'll use its steps to solve the Speedy Delivery problem.

In this example, your starting point is A. Therefore, you circle it, as shown here.

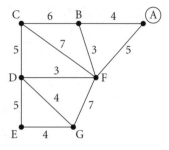

3. There are two nodes adjacent to A. What are they?

4. What are the two arcs from A to these nodes? Darken the arc with the shorter time. Circle the node at the endpoint of this arc.

5. List all nodes adjacent to the first two circled nodes.

6. List all paths starting at A that end at one of the uncircled adjacent nodes. Find the time from node A for each of these paths.

7. Circle the node and darken the arc that would create the shortest path to an uncircled adjacent node. What node did you circle? What arc did you darken?

Path AF and path ABF both lead to the uncircled node F. Choose AF over ABF because the time to node F is shorter. You should not choose path BF, because it is too long. Therefore, cross off arc BF on your graph.

8. List the uncircled nodes adjacent to the nodes you have already circled.

9. What are all the paths starting at A that lead to the uncircled adjacent nodes you identified in question #8 for which all of the nodes except the final node are circled and no node is used more than once? List them and the total time for each of these paths.

10. Circle the node and darken the arc that would create the shortest path. What node did you circle? Which arc did you darken?

This is a summary of the steps of the algorithm:

   I. List the beginning node on the table and circle it on the graph.

  II. List the uncircled adjacent nodes, the one-arc paths, and the total time in a table.

 III. Identify the shortest total time. In the table, underline the adjacent node that created the shortest time and the path. On the graph, circle the node and darken the path.

  IV. The node that was just identified has become a circled node in the table. Enter it in the Circled Nodes column of the table.

   V. Repeat step II with the new circled node. Find the shortest total time of *all* of the paths in the table that use at most one new arc and that are *not already underlined* (already used) or are *not crossed off* (too long).

   VI. Cross off all other paths to the circled node that are longer than the path you identified as fastest in step V.

   VII. Repeat steps IV and V, with the new circled node included.

11. What is the time of the fastest route from A to B? What is the fastest route from A to F? What is the fastest route from A to D?

12. Here is a table of what you should have found so far. Check that its entries agree with your answers. After circling node D in the graph and underlining it in the table, repeat steps IV and V. Then continue with the algorithm until you have found the fastest path to node E.

Circled Nodes	Uncircled Adjacent Nodes	Path from A	Total Time
A	<u>B</u> <u>F</u>	<u>AB</u> <u>AF</u>	<u>4</u> <u>5</u>
B	<u>C</u> F	<u>ABC</u> ~~ABF~~	10 ~~7~~
F	C <u>D</u> <u>G</u>	~~AFC~~ <u>AFD</u> <u>AFG</u>	12 <u>8</u> 12
D			

According to the shortest-path algorithm, there are no uncircled nodes adjacent to C. This means that the shortest path from A to E does not pass through C. You should also notice that once the table includes a path to E, you are not finished until you have circled E. *Once E is circled,* you have determined the shortest path to it. It is not possible to find some other path later from A to E that is shorter.

13. What is the fastest route from A to C?

14. What is the fastest route from A to G? What is unusual about this answer?

15. What is the fastest route from A to E?

16. Ms. Bosco, the dispatcher, needs to have the driver return to headquarters. Should she tell the driver to take the same route or a different route in order to arrive at headquarters as quickly as possible?

17. One of the drivers radios from E to Ms. Bosco that he needs to pick up a package at destination B before he returns to headquarters. Go through the shortest-path algorithm to find the shortest path from E to B. Show your work on a network and fill in a table, as in the previous example.

Everyday Applications of Operations Research

U.S. Coast Guard Uses Simulation to Improve Routes and Schedules for Buoy Tenders

The U.S. Coast Guard (USCG) maintains tens of thousands of buoys to aid navigation. It has approximately 40 ships, called buoy tenders, that maintain the buoys. Each buoy tender services several hundred buoys in its geographical region each year. Annual servicing consists of inspecting the buoy and its mooring chain, recharging its batteries, and replacing burned-out lights. At regular intervals, the buoy itself is replaced. The buoy tender's captain must plan the ship's activities so that each buoy is serviced within a one-month service window.

In the early 1990s, the USCG Research and Development Center investigated the routing and scheduling of the buoy tenders. The solution that was developed separates the routing and scheduling portions of the problem. The method was tested by computer simulations that included emergencies and adverse weather conditions, and was found to be efficient enough to accommodate them.

➤ For more case studies and background, go to [**www.hsor.org**].

EXTENSION 1: EXPANDED NETWORK

Due to the success of its quick delivery times, Speedy Delivery has expanded operations. Headquarters is still located at node A, but some additional locations have been added to the delivery area.

At 3:30 P.M., the driver has just made a delivery at point H. Ms. Bosco contacts the driver and informs him that the Pane Company at point N needs something picked up before 4 P.M. Refer to the network shown here and find the fastest route from point H to point N. Can the driver reach the company in time?

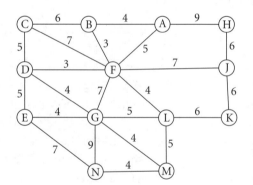

EXTENSION 2: ONE-WAY STREETS

In the network shown here, some of the routes are one-way streets (marked with arrows). A driver needs to go from point H to point N. Fill in the graph and the table according to the algorithm. Can the driver reach point N within 30 minutes?

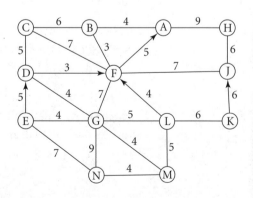

HOMEWORK

Tony Navarro drives a truck between the cities listed in the mileage table below. He would like to develop a schedule in which no city-to-city distance exceeds 100 miles. Using the mileages given in the table, draw a graph with nodes representing cities and arcs representing trips less than 100 miles. If your original graph has arcs that intersect, reposition the nodes and redraw the graph to eliminate the intersections.

	Ann Arbor	Detroit	Flint	Grand Rapids	Kalamazoo	Lansing
Ann Arbor	*	51	56	146	101	76
Detroit	51	*	62	156	136	90
Flint	56	62	*	121	134	56
Grand Rapids	146	156	121	*	50	91
Kalamazoo	101	136	134	50	*	78
Lansing	76	90	56	91	78	*

PROJECT 1: DIJKSTRA'S ALGORITHM

Learn as much as you can about E. W. Dijkstra's development of the shortest-path algorithm that carries his name. Write a brief report summarizing what you have learned.

PROJECT 2: ROUTING AT WORK

Visit the local post office, department of public works, delivery company, or any other institution that uses routes, and learn what problems they face in the construction and scheduling of routes. Write a brief report summarizing what you have learned.

PROJECT 3: A PRACTICAL ROAD MAP

Choose a state. Create a network showing the five largest cities and the major highways connecting them. Research the driving times or distances between the cities and label them on your map. Choose one of the cities on your map and, using Dijkstra's Algorithm, find the shortest path from the city you selected to each of the other cities on your map.

1

Service Woes at Speedy Delivery

In this lesson, students will study the basic elements of a network and solve a shortest-route problem. In routing applications, the optimal solution typically minimizes distance, time, or cost.

OBJECTIVES

- learn the meanings of nodes, arcs, paths, and numeric labels (distances, travel times, costs) associated with arcs and paths
- develop their mathematical reasoning skills
- learn to use an algorithm to find the optimal solution to a problem

INITIATING THE ACTIVITY

We recommend introducing the lesson by opening a discussion with your students about the shortest route from home to school. Some prompts you might use include, What criteria might be used to define "shortest"? Does "shortest" always refer to distance? Is the best route always the shortest route? (e.g., imagine walking home late at night.)

While students may determine the shortest time in the activity through a "guess and check" process, the algorithm developed in this module, Dijkstra's Algorithm, is a reliable method of finding the shortest path. Following the algorithm takes some practice. We recommend that you work out the problems ahead of time so that you can guide students later.

You may want to point out to students that although each node (except the starting point) first appears in the table as an uncircled adjacent node, each node on the shortest path will eventually also appear as a circled node. Conversely, some circled nodes may not end up on the shortest path.

Some students may have difficulty with question #6 (step VI of the algorithm). Make sure students understand that they need only to list paths in which no node is used more than once. To help students see that AF is the shortest path, explain that, with node A as a starting point, one should first choose path AB, because this path is shorter than AF; hence, node B becomes the next circled node. It may then occur to some students that BF is the shortest path. Although both paths ABF and AF end at the same node, you want only the path that is the shortest. Therefore, path ABF would never be taken in the network. We indicate this in the drawing by crossing off arc BF.

When C is circled, there is no need to repeat step II because no adjacent nodes are added. In addition, no new paths have been created to the uncircled nodes E and G, so the entries in the table do not need to be updated. In this example, the fastest path from A to E is found last, which will not always be the case.

SOLUTIONS

Activity

1. 10 minutes

2. Answers will vary.

3. B, F

4. AB, AF

5. C, F

6. ABC, 10; ABF, 7; AF, 5

7. F, AF

8. C, D, G

9. ABC, 10; AFD, 8; AFG, 12; AFC, 12

10. D, FD

11. AB, 4; AF, 5; AFD, 8

Filling in the table:

Circled Nodes	Uncircled Adjacent Nodes	Path from A	Total Time
D	C	AFDC	13
	E	AFDE	13
	G	AFDG	12

12. Equivalent alternatives:

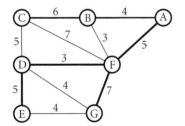

Circled Nodes	Uncircled Adjacent Nodes	Path from A	Total Time
A	B	AB	4
	F	AF	5
B	C	ABC	10
	F	~~ABF~~	~~7~~
F	C	~~AFC~~	~~12~~
	D	AFD	8
	G	AFG	12
D	C	~~AFDC~~	~~13~~
	E	AFDE	13
	G	~~AFDG~~	~~12~~
C	None		
G	E	AFGE	16

Circled Nodes	Uncircled Adjacent Nodes	Path from A	Total Time
A	B	AB	4
	F	AF	5
B	C	ABC	10
	F	~~ABF~~	~~7~~
F	C	~~AFC~~	~~12~~
	D	AFD	8
	G	AFG	12
D	C	~~AFDC~~	~~13~~
	E	AFDE	13
	G	AFDG	12
C	None		
G	E	AFGE	16

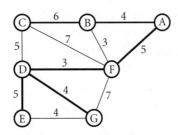

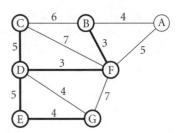

In the algorithm, at the time that node G was added to the list of circled nodes, there were two arcs or path choices that were equivalent. Either arc FG or arc DG could have been used to create a path of length 12 from node A to node G. The two diagrams represent the two alternative selections. The optimal path from A to E is still unique but need not be.

13. ABC

14. There are two equivalent paths, AFDG and AFG.

15. AFDE

16. Yes; the shortest path from E to A will be the same as the shortest path from A to E in this case. You may want to ask students under what condition(s) this might not be true. Extension 2, with one-way streets in the network, is one example.

17. Drawing and table: Optimal path is EDFB, and its time is 11 minutes.

Circled Nodes	Uncircled Adjacent Nodes	Path from A	Total Time
E	<u>D</u> <u>G</u>	EG ED	4 5
G	D F	~~EGD~~ ~~EGF~~	~~8~~ ~~11~~
D	<u>C</u> <u>F</u>	EDC EDF	10 8
F	C <u>B</u> A	~~EDFC~~ EDFB EDFA	~~15~~ 11 13
C	B	EDCB	16

Extension 1: Expanded Network

Yes. The shortest path is HJFLMN, and the time from H to N is 26 minutes.

Extension 2: One-way Streets

Yes. He can get from H to N in 28 minutes by path HJFGMN. Note: Due to the one-way streets, some nodes that look adjacent are not really adjacent; e.g., F is not adjacent to A.

➤ For complete solutions, go to [**www.hsor.org**].

Homework

One possible graph:

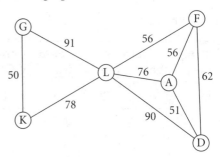

Activity 1: Service Woes at Speedy Delivery

The Speedy Delivery Company Network

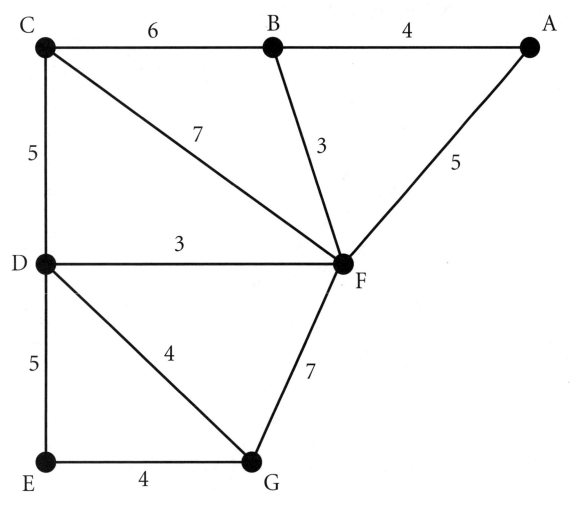

Circled Nodes	Uncircled Adjacent Nodes	Path from A	Total Time

Activity 1: Service Woes at Speedy Delivery

Speedy Delivery
Extension 1: Expanded Network

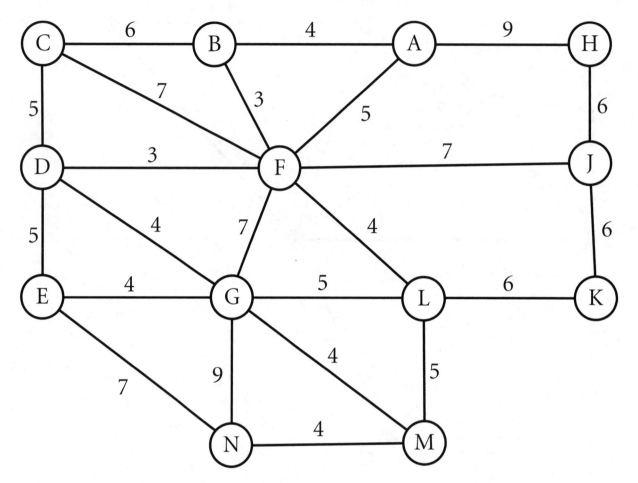

Circled Nodes	Uncircled Adjacent Nodes	Path from H	Total Time

ACTIVITY 2

Short Circuit Travel Agency

Steve Isaac works in the human resources department of an engineering company in the Washington, D.C., area. His job is to recruit top graduates from Carnegie-Mellon University in Pittsburgh, PA; Northwestern University near Chicago, IL; Washington University in St. Louis, MO; and Georgia Tech in Atlanta, GA. He plans to visit them all within a four-day period and then return to Washington, D.C. Shown below is a map indicating the cheapest available one-way fares between every possible pair of cities in either direction. Steve's travel agent at Short Circuit Travel Agency wants to determine the lowest total travel cost for his trip. This scenario is a variation of the classic traveling salesman problem.

This globe shows shipping routes, which follow the shortest routes between points on the globe.

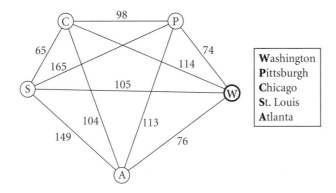

| Washington |
| Pittsburgh |
| Chicago |
| St. Louis |
| Atlanta |

This figure is a network, and the cities are the nodes. Steve's travel agent, Karl Rosiak, needs to find a route starting at W, passing through every other node in the network exactly once, and returning to W. Any route that visits every node exactly once and ends at the starting point is called a **Hamiltonian circuit.**

One method for finding the cheapest route is to list every possible Hamiltonian circuit and then compare the total cost of each. That way you can be sure that you have considered every possible circuit.

Considering All Circuits

1. Copy this table, which is a systematic list of all possible circuits that start with W and end at W. Record the circuit sequence.

Circuit #	Start	1	2	3	4	Return	Circuit Sequence	Total Cost
1	W	P	C	S	A	W	WPCSAW	74 + 98 + 65 + 149 + 76 = 462
2	W	P	C	A	S	W	WPCASW	
3	W	P	S	A	C	W	WPSACW	
4	W	P	S	C	A	W	WPSCAW	
5	W	P	A	S	C	W	WPASCW	
6	W	P	A	C	S	W	WPACSW	
7	W	C	P	S	A	W	WCPSAW	
8	W	C				W		
9	W	C				W		
10	W	C				W		
11	W	C				W		
12	W	C				W		
13								
14								
⋮								

2. How many possible circuits are there?

3. Among the circuits that start with WC, you should have identified circuit WCSAPW. If you traveled this circuit in reverse, what circuit would it be? Where in the list does this circuit already appear? Find the circuits that appear more than once in the list, and cross off the duplicates.

4. How many unique circuits remain?

5. How does the number of unique circuits compare to the total number of circuits?

6. The total cost of circuit 1 is $462, and its calculation is shown in the table. Calculate the cost for each of the remaining unique circuits and record it in the table.

7. Which circuit is the cheapest? What is its cost?

The Brute-force Method

The method you just used is sometimes called the "brute-force" method because it involves trying *every* unique circuit. Unfortunately, the larger the number of nodes, the larger the number of unique circuits. Let's begin developing a formula that represents this relationship.

8. How many nodes can you travel to directly from W?

9. After you have chosen the second node in your circuit, how many choices are there for the third node? Continuing this approach, how many choices are available for the fourth node in a circuit? For the fifth node?

10. Based on the answers to question #9, how many possible circuits are there?

11. Based on the answers to questions #4 and #10, what fraction of circuits is unique? Why does this fraction make sense?

Writing a Formula

Suppose n is the number of nodes in the complete network.

12. Write a general formula in terms of n for the total number of circuits that can be created starting at point W.

13. Recall that we discovered that some of the circuits were duplicates. Adjust the formula to account for these duplicates.

14. Suppose there were 6 cities. Use the formula to calculate the number of unique circuits. If there were 7 cities, how many unique circuits would there be? If there were 21 cities, how many unique circuits would there be?

15. What does your answer to question #14 suggest about the brute-force method?

16. The fastest high-speed computers can do approximately 1 trillion $\left(1 \times 10^{12}\right)$ computations per second. Assume that the construction of a 21-node path requires 100 computations. Using the brute-force method, how long would it take the computer to accomplish this task with 21 cities?

Expanding the Circuit

A visit to Ohio State has just been added to Steve's trip.

17. The diagram on the next page includes Columbus, OH, in the network, labeled with the travel costs to each of the other cities in the network. Start with the optimal solution to the original problem.

Suppose Karl wants to modify the original best route, WSCPAW, by having Steve fly first to Columbus and then on to St. Louis, instead of flying directly from Washington to St. Louis. What is the name of this new route?

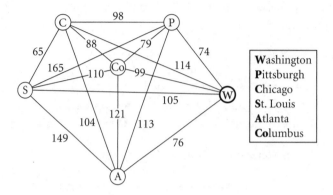

18. What is the cost of the new route?

The Nearest-neighbor Algorithm

Operations researchers have developed better algorithms for finding low-cost routes through a network. One method is called the **nearest-neighbor algorithm.**

The Nearest-neighbor Algorithm

I. Choose a node as your starting point.

II. From that starting node, travel to the node for which the fare is the cheapest. We call this node the "nearest neighbor." If there is a tie, choose one arbitrarily.

III. Repeat the process, one node at a time, traveling to nodes that have not yet been visited. Continue this process until all nodes have been visited.

IV. Complete a Hamiltonian circuit by returning to the starting point.

V. Calculate the cost of the circuit.

19. Use the nearest-neighbor algorithm to find a cost-efficient route for Steve's trip, starting and ending in Washington, D.C., and including a visit to Columbus. What is its total cost?

20. Why does using the nearest-neighbor algorithm make more sense than using the brute-force method in this case?

21. Will the nearest-neighbor algorithm always give a good route? Why or why not?

Everyday Applications of Operations Research

Tulsa Schools Save Money and Shorten Students' Travel Time

Transportation routes for special-education students require that the students be picked up at home and dropped off at selected schools that meet their specific educational needs.

In Tulsa, Oklahoma, the public school system determined bus routes for about 850 special-education students in 66 schools. Special-education students within the same neighborhood often attend different schools, and school hours vary from school to school. Many students also need to ride special buses that accommodate their disabilities.

Route planners were constrained by a 45-minute travel time limit for each student. Using routing methods, the Tulsa public school system was able to save time and money. Based on sample studies, the school system estimated that the routing algorithms reduced miles traveled by almost 11% and reduced time spent en route by almost 16%. It also estimated savings of $50,000 to $100,000.

➤ For more case studies and background, go to www.hsor.org .

EXTENSION 1: REPETITIVE NEAREST-NEIGHBOR ALGORITHM

One method that improves upon the nearest-neighbor algorithm applies the nearest-neighbor algorithm to each node and finds the best option. This method is called the **repetitive nearest-neighbor algorithm.**

The Repetitive Nearest-neighbor Algorithm

I. Select any node as a starting point. Apply the nearest-neighbor algorithm from that node.

II. Calculate the cost of that circuit.

III. Repeat the process using each of the other nodes as the starting point.

IV. Choose the "best" Hamiltonian circuit.

We will use the repetitive nearest-neighbor algorithm to find a good solution to the recruiting circuit scenario that includes Columbus.

1. In the activity, you applied the nearest-neighbor algorithm, using Washington, D.C., as the starting point. The total was $531. Apply the nearest-neighbor algorithm starting at Pittsburgh. (This begins the process of the repetitive nearest-neighbor algorithm.) What route does the algorithm generate? What is its total cost?

2. Consider the circuit that you found in question #1. What equivalent circuit begins and ends in Washington? What is the cost of this circuit?

3. Continue applying the repetitive nearest-neighbor algorithm until each of the cities has been used as a starting point. Which circuit is cheapest? What equivalent circuit begins in Washington?

4. When would the repetitive nearest-neighbor method be useful?

Notice that if there are n nodes in the network, this algorithm requires using the nearest-neighbor algorithm n times.

5. We have used three different algorithms. Compare them by completing this table.

	Brute Force	Nearest Neighbor	Repetitive Nearest Neighbor
Strengths			
Weaknesses			
When to Use			

EXTENSION 2: TRIANGLE-INSERTION ALGORITHM

When you add a node to the original network, you have to re-evaluate your circuit. An algorithm that allows you to tackle this problem efficiently is called the **triangle-insertion algorithm.** To use this algorithm, you begin by linking the new node to two nodes in the original circuit that are directly linked to each other and then eliminate the direct link. You find the cost (or length) of the new circuit by adding the sum of the costs of the two new links to the original circuit cost and subtracting the cost of the direct link that was eliminated. The process is repeated for each possible pair of directly linked nodes in the original circuit, and the cheapest (or shortest) resulting circuit is selected.

Triangle-insertion Algorithm

I. Begin with a Hamiltonian circuit for the original problem.

II. Choose a pair of directly linked nodes, link the new node to each of them, and then eliminate the old link between them. Calculate the cost of the new circuit by adding the sum of the costs of the two new links to the cost of the original circuit and subtracting the cost of the old link they replace.

III. On the original network, repeat the process by inserting the new node between every pair of nodes that are directly linked in the original Hamiltonian circuit.

IV. Choose the new circuit with the shortest length.

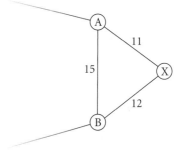

For example, you can add a new node, X to an existing circuit by linking it to two nodes, A and B, that are already linked in the existing circuit. In the new circuit, the edge AB will be replaced by the pair of edges AX and XB, as shown here. Now, if the old circuit had length L, the length of the new circuit is $L + (11 + 12 − 15)$.

The diagram below shows the addition of Columbus to the recruiting network. Columbus and the related travel costs are shown in bold in the diagram.

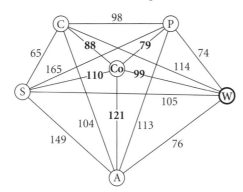

Apply the triangle-insertion algorithm to find the best route that includes Columbus.

PROJECT 1: RESEARCH ROUTE PLANNING

Use the library and/or the Internet to find out how routes are developed in different contexts. You can research school bus runs, pizza delivery routes, traveling sales routes, or routes of security personnel either in a building or over a region.

PROJECT 2: CONDUCT AN INTERVIEW

Work with a group. Select one of the applications suggested in Project 1, or select one of your own with your instructor's approval. Write a list of interview questions that you could ask to find out about that organization's method of route planning. Make an appointment and interview someone at the organization who is familiar with its method. Write a report, and present your report to the class. Be prepared to answer questions and to give an opinion of the route-planning method used by the organization.

PROJECT 3: DESIGN A BETTER ROUTING SYSTEM

Obtain information on bus routes in your district and create a route using the nearest-neighbor algorithm in this unit. Decide whether the route can be improved. Describe your findings.

PROJECT 4: OTHER ROUTING ALGORITHMS

Research other algorithms that have been used to address the classic traveling salesman problem.

Short Circuit Travel Agency

Students will be engaged in solving one of the classic routing problems: the traveling salesman problem (TSP).

OBJECTIVES

- use algorithms to find optimal or near-optimal solutions to routing problems involving Hamiltonian circuits
- analyze the efficiency of "brute-force" and heuristic algorithms by exploring the increase in the number of computations as the problem size grows and by developing a formula for the number of computations needed
- explore the trade-off between the efficiency and effectiveness of algorithms

INITIATING THE ACTIVITY

We suggest introducing the lesson by opening a class discussion concerning the best way to complete this list of errands in one trip:

- mailing letters at the post office
- making a deposit at the bank
- renting a movie at the video store
- purchasing items at the grocery store

Assuming you have a limited amount of time, what issues would you consider in deciding the order in which to complete the errands? In addition to distances and travel times, students might raise issues of convenience and timing. (If you walk, carrying groceries becomes an important consideration. If finding parking at each location takes time, there is a trade-off between driving and walking between locations. If the post office closes early, it might have to be the first stop regardless of other considerations.)

GUIDING THE ACTIVITY

Success as a travel agent requires finding convenient routes and competitive fares for your clients. Online booking systems also use algorithms to find the cheapest and best routes.

In a 5-node network, beginning at a specified node, there will be 24 Hamiltonian circuits that end at the same node. However, 12 of the 24 are redundant, because they merely traverse the network in the opposite direction. Cost, mileage, and similar factors would be unaffected by the reverse order. Thus, in questions #1 and #2, students are asked to generate all 24 circuits beginning and ending in Washington. Then, the intent of questions #3 through #5 is for students to discover that half of the circuits are redundant.

In general, the optimal solution to a TSP need not be a Hamiltonian circuit, but in this case it is. For a graph with just the three nodes S, C, and P, the optimal route starting at S goes from S to C to P and then retraces the route backward rather than flying directly from P to S. The triangle inequality does not hold for these data, which is not surprising for airline ticket prices.

The brute-force method always uncovers the optimal Hamiltonian circuit. Questions #6 and #7 ask students to compute the cost of each unique (nonredundant) circuit and identify the optimal (cheapest) route.

Questions #8 through #14 are intended to help students generalize a formula for the total number of unique circuits that start at a specific node, pass through every other node, and return to the starting point. If students are not familiar with factorials, you may want to consider introducing that concept and its notation at this point. In question #13, students should deduce that in a network containing n nodes there are $(n - 1)!$ Hamiltonian circuits that begin and end at a specified node. Then, in question #14 they should refer back to the result in question #5 to deduce that there are only $(n - 1)!/2$ unique circuits in the general case.

The point of questions #15 through #18 is twofold. First, we would like students to understand how problems of this nature quickly grow in size. Second, we introduce to students the idea that even the fastest computers using brute force cannot completely solve certain types of combinatoric problems, even ones that do not seem unduly large. The 21-city problem in question #17 does not seem large, but the total number of distinct routes is unmanageable in any practical way. In practice, there are numerous traveling salesman problems involving many more nodes. Therefore, while the brute-force method will always yield the optimal solution, it is not always feasible to use brute force. Question #18 demonstrates the impracticality of brute force. In addition, students get experience in converting units and develop a concrete understanding of a problem with very large numbers. A discussion about the "best solution" versus the "optimal solution" might ensue. In practice, companies are usually satisfied to find a very good solution, or an improved solution, rather than investing the time and money required to find the optimal solution.

Questions #20 through #23 are used to develop a solution strategy that does not use brute force. The addition of just one city to the 5-city network increases the number of nonredundant circuits that would have to be checked by brute force from 12 to 60. Hence, a different algorithm that might generate only a near-optimal solution is introduced. In Extensions 1 and 2, two additional algorithms are developed in the context of this same problem situation.

➤ For homework exercises and solutions, go to www.hsor.org .

SOLUTIONS

Activity

1.

Circuit #	Start	1	2	3	4	Return	Circuit Sequence	Total Cost
1	W	P	C	S	A	W	WPCSAW	74 + 98 + 65 + 149 + 76 = 462
2	W	P	C	A	S	W	WPCASW	74 + 98 + 104 + 149 + 105 = 530
3	W	P	S	A	C	W	WPSACW	74 + 165 + 149 + 104 + 114 = 606
4	W	P	S	C	A	W	WPSCAW	74 + 165 + 65 + 104 + 76 = 484
5	W	P	A	S	C	W	WPASCW	74 + 113 + 149 + 65 + 114 = 515
6	W	P	A	C	S	W	WPACSW	74 + 113 + 104 + 65 + 105 = 461
7	W	C	P	S	A	W	WCPSAW	114 + 98 + 165 + 149 + 76 = 602
8	W	C	P	A	S	W	WCPASW	114 + 98 + 113 + 149 + 105 = 579
9	W	C	S	P	A	W	WCSPAW	114 + 65 + 165 + 113 + 76 = 533
10	W	C	S	A	P	W	~~WCSAPW~~	
11	W	C	A	S	P	W	~~WCASPW~~	
12	W	C	A	P	S	W	WCAPSW	114 + 104 + 113 + 165 + 105 = 601
13	W	S	C	P	A	W	WSCPAW	105 + 65 + 98 + 113 + 76 = 457
14	W	S	C	A	P	W	~~WSCAPW~~	
15	W	S	P	C	A	W	WSPCAW	105 + 165 + 98 + 104 + 76 = 548
16	W	S	P	A	C	W	~~WSPACW~~	
17	W	S	A	P	C	W	~~WSAPCW~~	
18	W	S	A	C	P	W	~~WSACPW~~	
19	W	A	P	C	S	W	~~WAPCSW~~	
20	W	A	P	S	C	W	~~WAPSCW~~	
21	W	A	C	P	S	W	~~WACPSW~~	
22	W	A	C	S	P	W	~~WACSPW~~	
23	W	A	S	P	C	W	~~WASPCW~~	
24	W	A	S	C	P	W	~~WASCPW~~	

2. 24

3. WPASCW, which already appears as circuit 5

4. 12

5. It is one-half of the total number.

6. See the preceding table.

7. WSCPAW is cheapest, with a cost of $457.

8. 4

9. 3; 2, then 1

10. $4 \times 3 \times 2 \times 1 = 24$

11. One-half; each circuit can be traveled in two directions.

12. $(n - 1)(n - 2)(n - 3) \ldots (1) = (n - 1)!$

13. $(n - 1)!/2$

14. $(5)(4)(3)(2)(1)/2 = 60$; $(6)(5)(4)(3)(2)(1)/2 = 360$; $20!/2$, which is approximately 1.22×10^{18}

15. Brute force is not a realistic method when the number of nodes is large.

16. $1.22 \times 10^{18} \times 10^2$ computations$/(1 \times 10^{12} \times 60 \times 60$ min $\times 24)$, or more than 1400 days if run continuously. Not even the largest corporations in the world would pay for that much mainframe computer time to solve one fairly small problem.

17. WCoSCPAW

18. $561

19. WPCoCSAW. In this case, the starting point is not arbitrary, because the trip *must* begin and end in Washington, D.C.; $531.

20. With the addition of Columbus, the number of distinct routes increases from 12 to 60. This point was made in question #15.

21. It is likely to give a good route. However, it may miss the best route if the first leg of that route is inefficient.

Extension 1: Repetitive Nearest-neighbor Algorithm

1. Start at P: PWACSCoP = $74 + 76 + 104 + 65 + 110 + 79 = 508$

2. WACSCoPW; its cost is the same, $508. A circuit goes into and out of each city. Therefore, the starting point within a given circuit has no effect on the total cost.

3. Start at C: CSWPCoAC = $65 + 105 + 74 + 79 + 121 + 104 = 548$

   Start at A: AWPCoCSA = $76 + 74 + 79 + 88 + 65 + 149 = 531$

   Start at S: SCCoPWAS = $65 + 88 + 79 + 74 + 76 + 149 = 531$

   Start at Co: CoPWACSCo = $79 + 74 + 76 + 104 + 65 + 110 = 508$. This is identical to the circuit found by starting the nearest-neighbor algorithm at Pittsburgh. The cheapest circuit found by starting the algorithm at either Pittsburgh or Columbus translates to WACSCoPW and costs $508.

4. When you need a very good solution quickly, the repetitive nearest-neighbor algorithm will usually provide a good solution. In this case, the nearest-neighbor algorithm yielded a solution that cost $531. The repetitive nearest-neighbor algorithm reduced the cost from $531 to $508, a 4.3% reduction. The $23 savings is not large, but if you were a routing company, cutting daily transportation costs by 4.3% is significant and worth the extra time and effort. The optimal solution (by brute force) is $457, a 10% reduction from the repetitive nearest-neighbor solution. But for problems with many nodes, the extra computer time would be prohibitive.

5. *See table at bottom of next page.*

Extension 2: Triangle-insertion Algorithm

Different options for inserting Columbus into the circuit:

Insert Co between W and S: WCoSCPAW = $457 + ($99 + $110 − $105) = $561.

Insert Co between S and C: WSCoCPAW = $457 + ($110 + $88 − $65) = $590.

Insert Co between C and P: WSCCoPAW = $457 + ($88 + $79 − $98) = $526 (best insertion solution).

Insert Co between P and A: WSCPCoAW = $457 + ($79 + $121 − $113) = $544.

Insert Co between A and W: WSCPACoW = $457 + ($121 + $99 − $76) = $601.

Notice that there are significant differences between the alternative insertion points. Even the second-best insertion point is $18 or 3% worse than the best solution found by the triangle-insertion algorithm. Recall that the repetitive nearest-neighbor algorithm in Extension 1 found a solution that cost $508. Thus, the repetitive nearest-neighbor algorithm provided the better solution in this case.

PROJECTS

The first three projects are designed to give students some field experience. The last project requires research in the library or on the Internet.

Extension 1, #5

	Brute Force	Nearest Neighbor	Repetitive Nearest Neighbor
Strengths	Always produces an optimal solution.	Very quick. Can provide a quick result, even with many nodes.	Quicker than brute force and likely to provide a better solution than the nearest-neighbor algorithm. Solution is likely to be near optimal and is independent of the starting node.
Weaknesses	Time-consuming and may not be practical when there are many nodes.	Solution may be poor relative to the optimal because it is dependent upon the selection of the node that begins the algorithm.	Near optimal, but not necessarily optimal.
When to Use	When costs are extremely important, and the network is of modest size, saving even a small amount is critical.	When a route is needed immediately.	When you need a very good solution quickly.

Activity 2: Short Circuit Travel Agency

The Recruiting Circuit

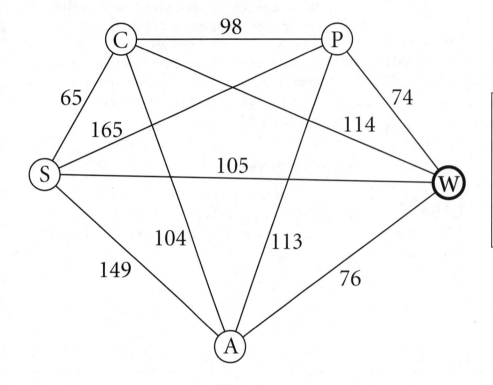

Washington
Pittsburgh
Chicago
St. Louis
Atlanta

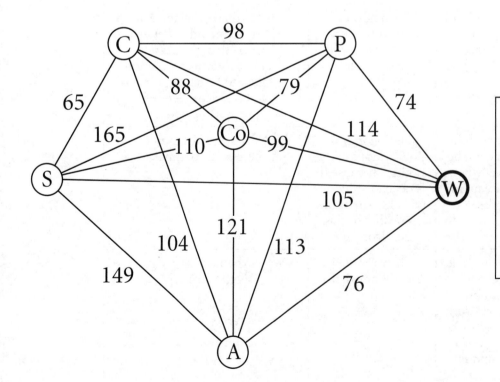

Washington
Pittsburgh
Chicago
St. Louis
Atlanta
Columbus

Linear Programming

BACKGROUND

Linear programming is an operations research technique that has been used to solve a wide variety of practical problems. Manufacturers produce an array of products with different prices, production costs, and profit margins. Their ability to maximize profits is limited, or constrained, by their machine, production line, and assembly line capacity, as well as by the size of their workforce. Manufacturers are also limited by marketing constraints on how much of a specific product they can sell and how much of it they need to produce in specific sizes, shapes, and colors to meet customer demand. Linear programming assists managers in making complex product-mix decisions in the presence of these constraints.

The founder of linear programming was American operations researcher George Dantzig. Between 1947 and 1949, he developed the foundation concepts for modeling and solving linear programming problems. The first problem Dantzig solved was a minimum-cost diet problem that involved the solution of 9 equations (nutritional requirements) with 77 decision variables. The problem had been posed in 1945 by the American economist George Stigler (a 1982 Nobel laureate), who had found a good but non-optimal solution. At the time he claimed, "There does not appear to be any direct method of finding the minimum of a linear function subject to linear conditions." Dantzig, working with The National Bureau of Standards, supervised the solution of the diet problem, which took 120 person-days using hand-operated desk calculators. Today, a personal computer can solve this problem in less than a second. Some linear programming problems that are solved by operations researchers today have tens of thousands of constraints and hundreds of thousands of decision variables.

The principles of linear programming can also be applied to nonlinear programming (in which the objective function or some constraints are nonlinear functions) or to integer programming (in which only integer values are considered). These methods are collectively called **mathematical programming.**

References

Dantzig, George B. 1982. Reminiscences about the origins of linear programming. *Operations Research Letters* 1:43–48.

Singhal, J., G. R. Bitran, and S. Dasu. 2001. Production management. In *Encyclopedia of operations research and management science,* edited by S. I. Gass and C. M. Harris, 639–647. Boston: Kluwer.

CASE STUDY: McDONALD'S FRANCHISES

Companies often use mathematical programming to arrange employee work schedules. For example, in the fast-food industry, work shifts vary, employee availability changes from week to week, and personnel requirements are different for "rush" times than for average or slow times. The primary concern in fast-food restaurant scheduling is matching part-time employees' availability with coverage requirements throughout the day. As a result, fast-food restaurants generally do not have standard work shifts or work weeks. Thus, compiling a workable scheduling system posed a challenge for Al Boxley, the owner of four McDonald's franchises in the Cumberland, Maryland, area.

Implementing a mathematical programming system enabled Boxley to compile manageable schedules for his employees. Developing the schedule by hand used to take him more than eight hours. In the late 1980s, operations researchers Robert Love and James Hoey were able to replace Boxley's previously burdensome and time-consuming manual scheduling with a more efficient microcomputer-based employee scheduling system. Because this problem involved 3 work areas, 30 work shifts, and thousands of decision variables, its solution would have been difficult, if not impossible, without the use of a computer. With the new scheduling system, the task was completed in less than an hour and the schedules better matched the schedule preferences of the workforce.

References

Love, Robert R., and James M. Hoey. 1990. Management science improves fast-food operations. *Interfaces* 20(2):21–29.

For more case studies and background, go to **www.hsor.org**.

ACTIVITY 3

High Step Sports Shoe

The High Step Sports Shoe Corporation wants to maximize its profits. The company makes two styles of sport shoe, Airheads and Groundeds. The company earns $10 profit on each pair of Airheads and $8.50 profit on each pair of Groundeds. How many of each shoe should the company produce per week?

Linear programming is often used to decide how much of each product to produce in order to maximize profit. In developing a production plan, managers will often be constrained by limited resources such as number of workers, availability of raw materials, maximum demand for a product, and so forth. Quantities that can change (vary) and that managers are able to control are called **decision variables.**

Decision Variables

1. What are the decision variables that the managers at High Step Sports Shoe must consider?

The Objective Function

2. The goal is to make the most money, or to **maximize** profits. Use a letter to represent each of the decision variables you identified in question #1, and write a function to model the profit earned from the manufacture of Airheads and Groundeds. This function is called the **objective function** because it models the goal, or objective.

The steps in manufacturing the shoes include cutting the materials on a machine and having workers assemble the pieces into a pair of shoes. The number of machines, workers, and factory operating hours limit the number of pairs of shoes that the company can make in a given time period. These limits are called **constraints.**

The High Step Sports Shoe Corporation has these constraints: There are 6 machines that are used to cut the materials. Each pair of Airheads requires 3 minutes of cutting time, while each pair of Groundeds requires 2 minutes. There are 850 workers who assemble the shoes. It takes a single worker 7 hours to assemble a pair of Airheads and 8 hours to assemble a pair of Groundeds. The assembly plant operates 40 hours per 5-day work week. Also, each cutting machine is operated only 50 minutes per hour to allow for routine maintenance.

Does This Line Ever Move? © 2005 Key Curriculum Press   **29**

3. How many minutes of work can 6 machines do in a 40-hour work week?

4. Use the decision variables to write an inequality to represent a constraint that is based on the limited time the cutting machines operate each week.

5. How many hours of work can 850 assembly workers do in a week?

6. Write an inequality to model a constraint based upon the limited number of worker hours available for shoe assembly each week.

7. Could the number of pairs of shoes of each style that are produced each week ever be negative? Could it be zero? Why or why not?

This machine uses a mold to make insoles for sports shoes.

8. Write two additional constraints based on your answers to questions #7.

Graphing the system of inequalities shows you the **feasible region** of the graph. The feasible region is the set of all feasible, or possible, solutions. The best solution is a point somewhere in this region.

9. A graph of the system of constraint inequalities is shown here. The letters *A* (for Airheads) and *G* (for Groundeds) were used to represent the decision variables. Label the graph of each straight line, including each coordinate axis, with its equation. The four straight lines in the graph intersect in six distinct points. Label each point of intersection with its coordinates.

10. What does each pair of coordinates represent?

11. Which of the six points of intersection satisfy all of the constraints? These points are called **feasible corner points.**

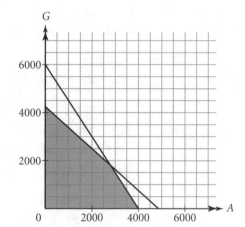

12. What does the shaded region of the graph represent?

Values of *A* and *G* that satisfy each of the constraint inequalities are considered feasible, or suitable, and the set of all such feasible points is the feasible region.

Of all the feasible points, one will give the maximum profit. The process of determining this best solution is called **optimization,** and the solution itself is called the **optimal solution.** To determine the optimal solution, you could try all of the feasible pairs of values.

13. How many feasible points are there?

We will investigate this problem further, in order to devise a more realistic strategy.

14. Pick three points in the interior of the feasible region. List the corresponding values of A and G in a table like this one, and evaluate the profit P for each point selected.

A (Airheads)	G (Groundeds)	$P = 10A + 8.5G$
1000	3000	$35500 = 10(1000) + 8.5(3000)$

Compare your answers with those of other students, and see which point has the most profit.

15. Now test each feasible corner point. Enter these values of A, G, and P in another table.

16. Which point from either table yields the largest profit?

17. What do the coordinates of this point represent?

This problem is an example of the **corner principle,** which states that an optimal solution to a linear programming problem always occurs at a corner point of the feasible region.

18. To see why the corner principle works, turn the profit equation around to yield $10A + 8.5G = P$, and substitute different quantities for the profit P (like \$30,000, \$35,000, \$40,000, \$45,000, and \$50,000). Then solve each equation for G in terms of A. What is the slope of each line?

19. Draw each line on the graph of the feasible region on the previous page. What do you notice about all of these lines?

20. Where on the graph are the lines that represent larger profits?

21. Do any of these values of P have feasible values of A and G?

22. Why does it make sense that the optimal solution lies at a corner point of the feasible region?

Everyday Applications of Operations Research

Plywood Ponderosa de Mexico Optimizes Product Mix and Increases Profits

Plywood Ponderosa de Mexico produces 85 million square feet of plywood each year. The demand for the type of wood needed is seasonal. In the winter, plywood is needed primarily for furniture manufacturing. In the summer, the construction industry is the main consumer of plywood.

Plywood Ponderosa de Mexico has 3 suppliers and uses 4 grades of logs. It produces many different panel grades and thicknesses. This makes the product-mix formulation complex. There are over 90 variables and 45 constraints.

By using linear programming in the early 1980s, Plywood Ponderosa de Mexico found that its earlier production of thicker plywood was not the most profitable. The model showed that thinner grades gave the most profit, and the company was able to increase its profits by 20%.

➤ For more case studies and background, go to (**www.hsor.org**).

EXTENSION 1: MARKET RESTRICTIONS

High Step's archrival, On-the-Run, Inc., recently started a major advertising campaign. High Step conducted a market study indicating that due to the increased competition, it could not sell more than 2500 pairs of Airheads or 3000 pairs of Groundeds per week.

1. Using the variables *A* and *G*, write the new sales constraints. On a new set of axes, graph these two new constraints, together with the cutting and assembly constraints from the first part of the problem. Be sure to label each line.

2. Does the solution to the original problem still work for this new problem? Explain your reasoning.

3. The feasible region now has four new corner points. Find the profit for each corner point.

4. Which corner point yields the most profit? What is this profit?

5. If advertising can increase demand, which product (Airheads or Groundeds) should High Step advertise more heavily? Justify your answer analytically.

EXTENSION 2: BEYOND OPTIMALITY

Dr. Swoopes is High Step's vice president for planning. She is trying to do two things: decrease waste and increase profits. First, Dr. Swoopes wants to know how many hours of cutting and assembly time are needed to produce the Airheads and Groundeds that correspond to the optimal strategy for Extension 1.

1. Use the cutting-time constraint to determine the number of minutes of cutting time needed to produce the optimal number of each style of shoe. (Be sure to use the answer from question #4 in Extension 1). Is the cutting-time resource fully used?

2. Use the assembly-time constraint to determine the number of worker hours needed to assemble the optimal number of each style of shoe. (Be sure to use the answer from question #4 in Extension 1.) Is the assembly-time resource fully used?

3. Dr. Swoopes is considering several alternatives to either increase resources or obtain more advertising. She wishes to evaluate each of these activities separately.

   - increase cutting time by running each machine for an additional 2 hours per day
   - increase assembly time by hiring 50 more workers
   - increase demand for Airheads by 200 pairs per week

   Write a new constraint for each of these alternatives. How does each one affect the feasible region? Find the optimal corner point. Then determine which change would have the greatest impact on weekly profits.

HOMEWORK

1. **The Not-Whole Plywood Company.** In plywood manufacture, logs are cut into thin sheets called green veneer, which are dried and cut into different sizes. There are different grades of logs and different grades of veneer. Several sheets of veneer are then glued together and pressed in a hot press. In the final stage, the rough plywood is sawed into an exact size and polished. The result is plywood panels of different grades and thicknesses.

   You are the production manager of the Not-Whole Plywood Company. You have 48,000 sheets of grade A veneer and 60,000 sheets of grade B veneer for use in this week's production. Not-Whole manufactures two products: premium and regular plywood pieces. Each piece of premium requires 3 sheets of grade A and 2 sheets of grade B veneer and sells for $14. Each piece of regular uses one sheet of grade A and three sheets of grade B veneer and sells for $11. Each premium plywood sheet requires 2 minutes of polishing and 40 seconds of pressing. Each regular plywood sheet requires 1 minute of polishing and 30 seconds of pressing. You have 20 polishing machines and 5 presses. You can run each machine for 40 hours during the week.

a. Formulate the plywood production problem to maximize this week's revenues.
b. Find the optimal corner point and describe the optimal production plan. What is the profit the company can expect with this plan?
c. In your optimal plan, are the polishing machines and presses in use all 40 hours? Do you use up all the grade A and grade B veneer?

2. **General Post Farms.** The forecast for this year's U.S. farm production of soybeans is approximately 2.7 billion bushels, and production of corn is forecasted to be 9 billion bushels. The most recent wholesale price forecasts are for soybeans at $5.20/bushel and corn at $2.50/bushel.

You are in charge of planting a 1000-acre section of your family farm and are trying to decide how many acres of each crop to plant. In your experience with your farm, an acre planted with soybeans produces 40 bushels and an acre planted with corn produces 120 bushels. To maintain a stable income, you have signed long-term contracts with the ADP Processing Company to provide a minimum of 10,000 bushels of soybeans and 24,000 bushels of corn. Over the whole season, each acre of corn requires an average of 10 hours of work and each acre of soybeans requires only an average of 6 hours of labor. In total, you have only 7200 hours of available labor during the season.

a. Formulate the crop-planting problem to maximize this season's revenues.
b. Find the optimal corner point and determine the optimal planting plan.
c. How much additional revenue could you earn if there were 100 more hours of labor available? What is the value of each additional hour? Why?
d. Crop yields are heavily influenced by the amount of rain. The National Weather Service forecasts a season dryer than normal. If the forecast is accurate, you are concerned that the farm yields will be down by 15% for soybeans and 10% for corn. How would this change your formulation of the crop-planting problem?

PROJECT 1: ALLOCATE STUDY TIME FOR FINAL EXAMS

Final exams are coming up soon, and you are trying to make plans to study.

1. What decision variables are under your control? Write a statement in words that represents a decision variable.

2. What constraints do you face? Are they all less-than-or-equal-to constraints? Are some of them greater-than-or-equal-to constraints?

3. What is the ultimate objective that will guide your decisions? Are you trying to maximize something or minimize something? How does your grade point average going into the final exam affect your decisions?

4. Describe a hypothetical final-exam scenario, and write an objective function. Does each hour of studying in a particular course add a given number of points to the estimated final exam score? Is the final exam score a linear function of the number of hours spent studying? Would letter grades versus numeric grades make any difference in the formulation?

PROJECT 2: INTERVIEW A MANAGER OF A LOCAL BUILDING SUPPLY STORE

Visit a local building supply store to gather data on prices of plywood. Find out how the price varies by the size of the sheet and its thickness. Calculate whether or not the price is directly proportional to the size and/or thickness. Find out how many sheets of veneer go into a piece of plywood. What other factors influence the price of the different sheets of plywood?

ACTIVITY 3

High Step Sports Shoe

This is a "product-mix" problem, and it occurs whenever a company produces more than one item. Students will learn how to apply the solution of a system of linear inequalities to solve geometrically a real-world problem involving two decision variables.

OBJECTIVES

- convert statements in ordinary language into the language of mathematics
- connect the solution set of a system of linear inequalities, which constitutes the feasible region in this context, to a production plan that could be accomplished
- link an objective function graphically to the feasible region in order to determine a "best" solution
- understand the corner principle

TECHNOLOGY NOTES

With graphing calculators or graphing software, students can graph feasible regions quickly. Most graphing utilities can shade above or below the function, though they usually show a solid line or curve, regardless of the type of inequality. Vertical boundary lines can be approximated with a line of very steep slope.

In some cases, graphing several inequalities on the same screen makes the shading of the overlapping regions too dark, and the feasible region is not easily visible like the screen at right.

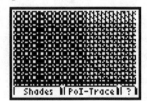

One way around this problem is to use reverse shading (i.e., shade the "wrong" side of the function). Then the feasible region is unshaded and its boundaries and interior are easy to see.

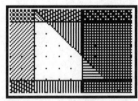

Some graphing programs also have an application just for graphing inequalities, and will automatically shade only the feasible region.

TEACHER NOTES

PRE-ACTIVITY: LEGO FURNITURE *(optional)*

As a warmup, you can use Lego® pieces to model production in a furniture factory with your students. Suppose a company produces only tables and chairs. A table is made of two large and two small pieces, while a chair is made of one large and two small pieces. The resources available are six large and eight small pieces.

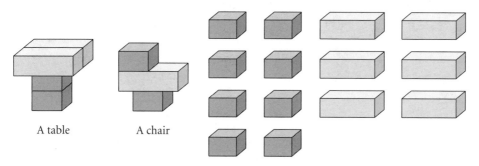

A table   A chair

Profit for a table is $16 and for a chair is $10. What product mix maximizes the company's profits using the available resources?

Guiding the Pre-activity

You might ask these questions beforehand: What information does a company need in order to make a choice about how many of each product to produce? How could the company collect this data? How often do you think the company should update its database about markets and products? Why?

You may want to have your students define decision variables that correspond to the number of chairs and tables produced. Then have them use the decision variables to define an objective function and write any constraints on the decision variables.

Pre-activity Solutions

If students start by trying to make as many tables as the available resources allow, they will find a solution of 3 tables with a profit of $48. However, making 3 tables does not allow them to make any chairs. This attempt at a solution is one corner point of the feasible region, (3, 0). Another corner point, (0, 4), represents making as many chairs as possible and produces a profit of $40. However, the optimal solution in this case occurs at the intersection of the other two constraints:

Let T = the number of tables and C = the number of chairs.

Then $P = 16T + 10C$ is the objective function to be maximized subject to the following constraints:

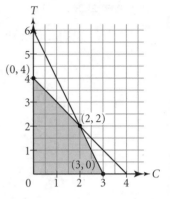

$2T + 1C \leq 6$ because only 6 large pieces are available, and a table requires two large pieces while a chair requires only one.

$2T + 2C \leq 8$ because only 8 small pieces are available, and tables and chairs each require 2 small pieces.

$T \geq 0$ and $C \geq 0$

This activity was adapted from Pendegraft, Norman. 1997. Lego of my simplex. *OR/MS Today* 24(1):128.

GUIDING THE ACTIVITY

Some students may question whether the decision variables must be integers because they represent the number of pairs of shoes produced. However, in this context the variable is really the number of pairs of shoes produced *per week,* or production *rates.* Therefore, they need not be integers. A pair of shoes may be only partially assembled when the workers go home for the weekend.

Stress the importance of the constraints and how constraints affect this manufacturing industry. For instance, you could ask, if the goal is to maximize profit, why not just manufacture a huge number of both kinds of shoes?

In discussing the formulation of this problem, introduce and use the language of linear programming. The variables in the problem are **decision variables,** and restrictions on the decision variables are **constraints.** The quantity to be optimized should be expressed as a function of the decision variables, and this function, because it mathematically describes the basic objective in the problem, is named the **objective function.**

When graphing inequalities, students may have difficulty determining which half-plane to use. In that case, they should be encouraged to use a representative test point. For equations that do not pass through the origin, (0, 0) is the simplest point to test. Because all of the constraints are of the less-than-or-equal-to or greater-than-or-equal-to types, their graphs include the line itself. Thus, the corresponding equations should be graphed as solid lines.

The lesson concludes by helping students form a rationale for the corner principle. Ask students to consider the possibility that the profit lines in the last part of the activity could be parallel to one of the boundary lines of the feasible region. In that case, every point on that boundary line represents an optimal solution. Determining a set of profit values for which this would be true is a nice real-world application of the slope of a straight line.

SOLUTIONS

Activity

1. The decision variables are the number of Airheads (A) and Groundeds (G) to produce each week.

2. Profit, $P = 10A + 8.5G$

3. $6 \times 50 \times 40 = 12000$ minutes

4. $3A + 2G \le 12000$

5. $850 \times 40 = 34000$ hours

6. $7A + 8G \le 34000$ (If students believe that the units for both constraints must be the same, this would be a good place to discuss equivalent equations.)

7. Zero is possible, but you cannot make a negative number of shoes.

8. $A \ge 0$ and $G \ge 0$

9.

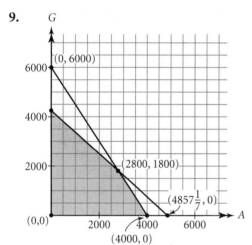

10. The intersection points are $(0, 4250)$, $(0, 0)$, $(4000, 0)$, $(2800, 1800)$, $(0, 6000)$, and $(4857\ 1/7, 0)$. The coordinates represent the number of pairs of each style of shoe produced.

11. Only $(0, 4250)$, $(0, 0)$, $(4000, 0)$, and $(2800, 1800)$

12. The set of all possible plans for producing Airheads and Groundeds that satisfy all of the constraints

13. There are infinitely many

14. Answers will vary, but each point should lie in the interior of the feasible region.

15. $10(0) + 8.5(4250) = 36125$

    $(10)(4000) + 8.5(0) = 40000$

    $10(2800) + 8.5(1800) = 43300$

16. $(2800, 1800)$

17. Produce 2800 Airheads per week and 1800 Groundeds per week, for a total profit of \$43,300.

18. $-10/8.5$, or $-20/17$

19. They are parallel.

20. The profit, (P), increases as the lines move farther up and to the right.

21. Yes: 30,000, 35,000, and 40,000.

22. If you visualize the parallel lines as a single moving line, as the line moves in the increasing P direction, the last point this line will touch before it leaves the feasible region entirely will be a corner point.

Extension 1: Market Restrictions

1. $A \le 2500$ and $G \le 3000$

2. No; it is now outside the feasible region.

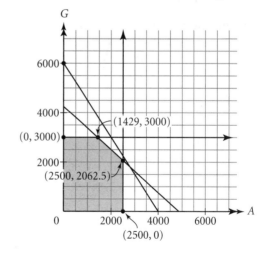

TEACHER NOTES

3.

A	G	10A + 8.5G
0	0	$0
0	3000	$25,500
1428 4/7	3000	$39,785.71
2500	2062.5	$42,531.25
2500	0	$25,000

4. (2500, 2062.5) for a profit of $42,531.25.

5. Airheads. When Airhead production is increased as Groundeds are decreased, profit goes up. Airheads make more money and take less time for a worker to put together.

Extension 2: Beyond Optimality

1. $3(2500) + 2(2062.5) = 11625$; no.

2. $7(2500) + 8(2062.5) = 34000$; yes.

3. Increasing the cutting-time constraint does not affect the solution, because the original line $3A + 2G = 12000$ and the new line $3A + 2G = 15000$ both lie entirely outside the feasible region.

Increasing the assembly time produces a new constraint, $7A + 8G \leq 36000$, which expands the feasible region slightly upward and to the right. Increasing the assembly time yields a maximum profit of $44,656.25 at (2500, 2312.5).

Increasing the demand for Airheads yields a maximum profit of $43,043.75 at (2700, 1887.5).

Increasing the available assembly time has the greatest impact on weekly profits.

HOMEWORK

1. The Not-Whole Plywood Company
  a. Formulation: P = number of pieces of premium and R = number of pieces of regular.
  Maximize: Profit = $14P + 11R$, subject to these constraints:

$3P + 1R \leq 48000$ (grade A sheets)

$2P + 3R \leq 60000$ (grade B sheets)

$2P + 1R \leq 48000$ minutes (polishing)

$40P + 30R \leq 720000$ seconds (pressing)

$P \geq 0$ and $R \geq 0$

  b. They should manufacture 6000 sheets of premium and 16,000 sheets of regular, for a profit of $260,000.

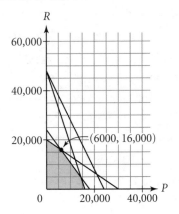

  c. The presses are used to the maximum but not the polishing equipment. Only 28,000 minutes of polishing are used. The entire grade B veneer is used but not all the grade A. There are 14,000 sheets of grade A veneer left over with the optimal plan. The polishing resource is so large relative to the other constraints that it does not affect the feasible region.

2. General Post Farms
  a. Formulation: Let S and C be the number of acres (*not* bushels) of soybeans and corn, respectively.
  Maximize $P = (5.2)40S + (2.5)120C = 208S + 300C$, subject to these constraints:

$S + C \leq 1000$ acres

$40S \geq 10000$ bushels

$120C \geq 24000$ bushels

$6S + 10C \leq 7200$ hours

$S \geq 0$ and $C \geq 0$

The 1000-acre constraint should not be written as an equality, because there may not be enough labor hours available to work the entire 1000 acres. Note that the constraints $S \geq 0$ and $C \geq 0$ are not really necessary, because this is already guaranteed by the constraints $40S \geq 10000$ and $120C \geq 24000$.

b. Optimal plan: $S = 700$ and $C = 300$. Total revenue = \$235,600.

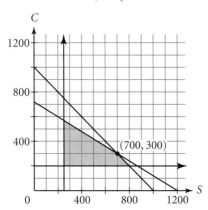

c. The added value is \$2300. $S = 675C = 325$. Total revenue = \$237,900. Each hour of the 100 hours is worth \$23.

d. New formulation: Remember that this change affects not just the objective function but also the requirements to meet the minimum contract commitments with ADP. Maximize $P = 176.8S + 270C$, subject to these constraints:

$$S + C \leq 1000$$
$$34S \geq 10000 \text{ bushels}$$
$$108C \geq 24000 \text{ bushels}$$
$$6S + 9C \leq 7200 \text{ hours}$$
$$S \geq 0 \text{ and } C \geq 0$$

PROJECTS

In the first project, students are challenged to think about decision variables, constraints, and an objective function in the context of studying for exams. The second project is intended to reinforce the applications of the lesson. It is connected to one of the homework examples and involves collecting data from a local business about different types of plywood. You may change the project, substituting other commodities, depending on the industries in your locale.

TEACHER NOTES

Activity 3: High Step Sports Shoe

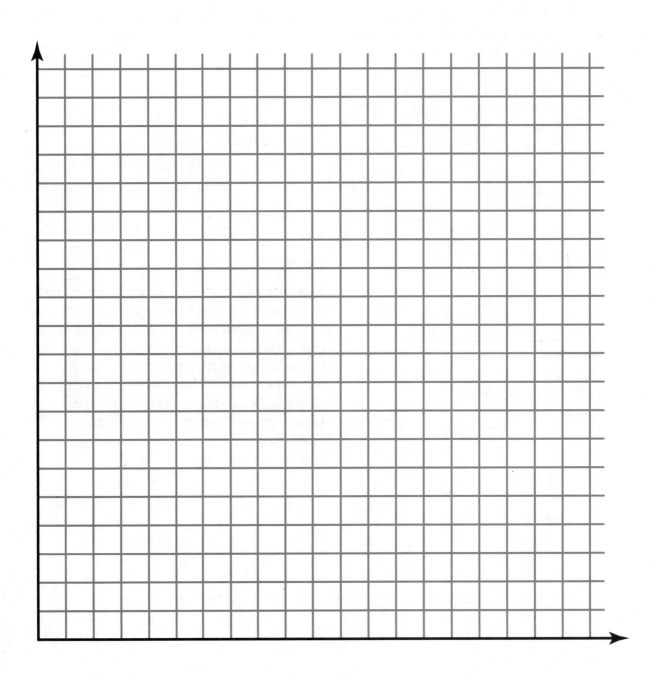

Does This Line Ever Move? © 2005 Key Curriculum Press

ACTIVITY 4

Trim Loss at the "Cutting Times"

Products such as paper, sheet metal, and photographic film are manufactured in very wide rolls that need to be cut into sets of smaller strips to meet specific product needs. Linear programming is used to determine what cutting patterns to use in order to minimize the amount of trim loss. The main constraints are set by customer demand.

A worker at a printing press checks the print quality on a newspaper pressrun.

THE TRIM-LOSS SCENARIO

The *Cutting Times* is a small local newspaper with its own printing press. The paper receives large 48-inch-wide rolls of newsprint, which must be cut into 25-inch-wide rolls for the ordinary pages in the paper and 21-inch-wide rolls for smaller inserts.

1. Draw models of each possible cutting pattern. Label each roll with its dimensions, and label any waste on each roll. Describe the two most efficient patterns for cutting the large rolls of newsprint to produce the smaller rolls needed. Be sure to include the number of inches of newsprint wasted for each pattern.

	Number of 25-inch Rolls	Number of 21-inch Rolls	Inches of Paper Wasted
Pattern A			
Pattern B			

2. For the Sunday edition of the *Times* right before Thanksgiving, the *Times* needs twenty 25-inch rolls for the ordinary part of the edition and fifty 21-inch rolls for the extra advertising inserts. How many of each pattern should be cut? Compare your recommendation with those of others in your group. Discuss the reasons for each recommendation.

3. When the *Times* decides how many of each pattern to cut, what should be considered?

4. How can the newspaper control the amount of waste?

5. Why would the newspaper want to control the amount of waste?

Decision Variables

The next step is to identify the decision variables that can be controlled. In this case, the decision is how many rolls to cut for each pattern.

Let x = the number of large rolls cut into two 21-inch rolls.

Let y = the number of large rolls cut into one 25-inch and one 21-inch roll.

6. Label your patterns in question #1 with x and y. These represent the quantity of each pattern. Write a function representing the amount of waste, w, in terms of the decision variables x and y. (Is the *Cutting Times* interested in minimizing or maximizing waste?)

Because w is a function of x and y and represents the newspaper's goal, or objective, the function you have just written is the objective function.

Constraints

Next, consider the paper requirements in relation to how the rolls are supplied.

7. When a large roll is cut into two 21-inch rolls, how many 25-inch rolls and how many 21-inch rolls are produced?

8. When a large roll is cut into one 25-inch roll and one 21-inch roll, how many 25-inch rolls and how many 21-inch rolls are produced?

9. If 5 rolls are cut using the pattern described in question #7, and 10 rolls are cut using the pattern in question #8, how many 25-inch rolls and how many 21-inch rolls are produced?

10. If x rolls are cut using the pattern described in question #7, and y rolls are cut using the pattern in question #8, how many 25-inch rolls and how many 21-inch rolls will be produced?

11. Use the paper requirements in question #1 to write two inequalities based on the number of 25-inch and 21-inch rolls that must be produced, in terms of x and y. Each of these inequalities represents a constraint that restricts the values of the decision variables.

The Feasible Region

12. Graph the system of constraints, using graph paper or a graphing program. One of the constraints restricts the possible values of only one of the decision variables. Which decision variable is this? In terms of the problem, why is this variable restricted in this way?

    Shade the region of the plane that satisfies all the constraints. The region you have just shaded, which contains all possible, or feasible, solutions, is called the feasible region. All possible solutions lie in the feasible region, so the best solution to the problem must lie somewhere in this region.

13. On your graph, plot the point that represents your recommendation from question #2. Is your recommendation a possible solution?

The Optimal Solution

Next, in order to identify the optimal solution, you will investigate the amount of waste produced by various solutions.

14. Choose two points in the feasible region and label them on the graph. In terms of the problem, discuss what these points represent.

15. Use the objective function to calculate the amount of waste $\left(w_1 \text{ and } w_2\right)$ produced for the points you selected.

    Set the expression that defines the objective function equal to w_1 and w_2 to obtain two equations. Graph these two equations on your graph of the feasible region.

16. What do you notice about these two graphs? How can you describe the location of the line that represents the smaller amount of waste?

17. Compare answers with your group. What region of the graph represents the least waste?

18. On your graph of the feasible region, draw the line that represents the least possible waste and still intersects the feasible region. At what point does this line intersect the feasible region? How would you describe the location of this point in the feasible region?

19. What do the coordinates mean in terms of the original problem?

20. What is the least amount of waste possible?

21. If there is a unique optimal solution to a linear programming problem, it must occur at one of the corner points of the feasible region. Explain why this is true.

Additional Constraints

So far, we have assumed that any extra 25-inch rolls that were cut could be stored for future use. However, a business might not always be able to do so.

22. If extra 25-inch rolls that are cut cannot be stored for future use, how must they be treated in the formulas?

23. Based on your answer to question #22, how many large rolls should be cut using pattern *y*?

24. How many 21-inch rolls and how many 25-inch rolls will this produce?

25. How many large rolls should then be cut using pattern *x*?

26. Which point in the feasible region represents this solution?

27. How much waste is generated with this solution?

Everyday Applications of Operations Research

Kendall Corporation Uses Linear Programming in Gauze-Slitting Operations

The Kendall Corporation produces health care products, such as bandages and sterile pads, from large beams of gauze. At its Augusta, Georgia, plant, between 18 and 20 slitting schedules per week are now determined by a linear programming computer program rather than developed informally by schedulers and operators. Each slitting combination specifies the width to which the gauze should be stretched, the combination of slit widths, and what portion of the 25,000 yards should be cut using this combination. In the early 1990s, the use of linear programming to determine the slitting schedules saved the Kendall Corporation $500,000 in its first full year of production under the new system.

➤ For more case studies and background, go to (**www.hsor.org**).

EXTENSION 1: ALL THINGS CONSIDERED

In the Activity, you considered any excess rolls of either size to be waste and, informally, found a new optimal solution that accounted for this waste. It is possible to include this aspect of the problem by introducing two additional variables.

Let E_{21} = the number of excess 21-inch rolls and E_{25} = the number of excess 25-inch rolls.

The consideration of these variables changes the constraints you wrote in question #11.

1. Write expressions for E_{21} and E_{25}.

2. How are these constraints different from those you found in question #11?

3. How must the objective function be changed to reflect the impact of this expanded definition of waste?

4. With this new formulation, what problem do you have in finding the optimal solution?

HOMEWORK

1. **Gauze solutions.** A health care corporation wants to cut large 55-inch-wide beams of gauze into 5-inch and 4-inch widths.

   a. List all the possible cutting patterns.
   b. What variable names could be assigned to represent the number of beams of gauze cut using each of the patterns?
   c. The company also wants to cut some of the 5-inch rolls of gauze into 3/4-inch rolls to be used in the manufacture of adhesive strips and 3-inch rolls to be used in the manufacture of square gauze pads. The company needs 120 thousand 3/4-inch rolls and ten thousand 3-inch rolls. How should the 5-inch rolls be cut to minimize waste?

2. **Art posters.** Mrs. Miller has 3 sheets of 36-by-24-inch poster board that she wants to cut into 12-by-9-inch and 5-by-6-inch pieces. She plans to have each of her 16 students sketch student portraits on the larger sheets and silhouettes on the smaller sheets. Make drawings of possible patterns. Find a pattern for cutting the sheets that produces at least 1 large piece of waste that can be used later in another project. Write a paragraph that explains why your pattern is the optimal solution. Be sure to include the number of each of the 2 patterns that could be cut from each of the 3 sheets.

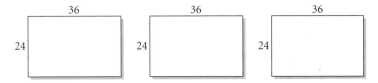

PROJECT 1: APPLICATIONS OF TRIM-LOSS CONCEPTS

What other industries might use the trim-loss concept to optimize their production? Research and describe their processes.

PROJECT 2: 101 USES FOR NEWSPRINT

What creative ways are there to use the waste from the newsprint that would not require storage or disposal?

PROJECT 3: EXPLAIN OPTIMIZATION

Discuss the term "optimization." Choose a scenario not discussed in class.
   a. What would you want to optimize?
   b. What constraints would you have to consider?
   c. What compromises would you have to make to find a solution to satisfy the most people?

Trim Loss at the "Cutting Times"

Students will learn how to apply the solution of a system of linear inequalities to solve geometrically a real-world problem involving two decision variables.

OBJECTIVES

- formulate a problem using mathematical expressions and interpret the solution
- graph and solve systems of inequalities
- make informed decisions while interpreting the data

Additional objectives for the extension are to analyze the impact from the addition of more constraints or more decision variables, which takes the problem beyond the two-dimensional coordinate system.

GUIDING THE ACTIVITY

The *Cutting Times* is an example of a trim-loss problem that occurs when a number of items are cut from a larger piece of material, sometimes producing waste. These items should be cut to minimize the amount of waste material while meeting a required amount for each item. Trim-loss problems are found in many industries, including the steel, paper, copper, meat, and clothing industries.

The pre-activity explores the combinatorics involved when patterns vary. No solutions are provided, because students should use their intuition to compare different possibilities. This pre-activity is intended to give students an appreciation of the many variables involved and the difficulty of using traditional heuristic methods for solving more complex problems.

PRE-ACTIVITY: WRAPPING PAPER

The Holiday Company makes a 60-inch roll of wrapping paper, which needs to be cut into 10-inch, 20-inch, 25-inch, and 30-inch widths for various customers. How many different cutting patterns can they use? (You can give each group approximately ten sticky notes. If a group finds a pattern that has not already been posted on the board, ask them to write it on the sticky note with the team number and put it on the board. After all the patterns have been posted, check and discuss each one in a whole-class format. Make a larger list by combining all the patterns.)

Which pattern works best for each of these customers? (Give each group one of the following customers to service. Have the group decide which pattern or patterns meet the needs of this customer while producing the least waste and present their decision to the class with a justification.)

Customer Number	Number of 10-inch Rolls	Number of 20-inch Rolls	Number of 25-inch Rolls	Number of 30-inch Rolls
1	107	109		110
2	80		25	
3	89			29
4			40	51
5	63	20		
6	58	40		
7		27		8
8	120		15	
9	78	15		33
10	59	28		23

GUIDING THE ACTIVITY

After students have read through the end of the first question in the *Cutting Times* activity, they should be able to list the specific components of the problem. Students should discover the patterns illustrated here.

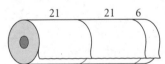
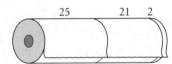

An alternative approach to finding patterns is to give students physical models (e.g., discarded tubes of paper towel rolls) to cut and label.

If students hit upon the optimal solution (15, 20) intuitively in question #2, encourage them to work out the solution to check their answer. Extension 1 will provide a more challenging problem that is not easy to solve intuitively.

Continually emphasize the appropriate vocabulary: decision variables, objective function, constraints, feasible region, optimal solution.

Emphasize that x and y do *not* represent the number of 21-inch and 25-inch rolls, respectively. Rather, x represents the number of large rolls cut using the first pattern, and y represents the number of large rolls cut using the second pattern. Also discuss the assumed constraints $x \geq 0$ and $y \geq 0$.

After question #13, you might choose to graph the feasible region on a transparency and have a few students plot their recommendations from question #2. As a class, determine whether the point lies within the feasible region and use the objective function to determine the amount of waste for the plotted point. Now have students independently choose two points in the feasible region. As a demonstration for question #13, plot points that students chose, calculate the waste for each point, set the expression that defines the objective function equal to the amount of waste calculated, and graph the resulting line on the graph showing the feasible region. Each such line is called a **line of constant waste,** because the waste is the same for each point on the line. Students may have to graph more than two equations $\left(w_1 = 6x + 2y \text{ and} \right.$ $\left. w_2 = 6x + 2y \right)$ before they see the pattern of parallel lines in question #16.

In setting up the series of graphs, it is a good idea to include lines passing through each of the corner points of the feasible region. Observing a series of parallel lines will enhance students' discovery of the optimal solution in questions #14–20.

Alternatively, a series of lines of constant waste may be animated using a graphing calculator. For Texas Instruments graphing calculators, use the Cycle Pic command. Store a series of graphs showing the feasible region and lines in the form $6x + 2y = w$, for values of w differing by a constant.

In this problem all of the decision variables are full rolls of paper and must take integer values. Therefore, the feasible set for this problem really includes lattice points only within the constrained area. The graphical logic of creating a series of parallel lines and moving them in the direction of decreasing waste will still identify the optimal *integer* solution. It will be the last lattice point reached before leaving the feasible region.

Note that the optimal solution need not occur at (or even near) one of the corners of the feasible region in such a problem, as shown here.

Ask students to summarize their findings and to compare the optimal solutions with and without the additional constraints.

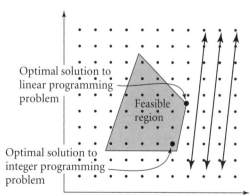

Optimal solution to linear programming problem

Feasible region

Optimal solution to integer programming problem

SOLUTIONS

Activity

1.

	Number of 25-inch Rolls	Number of 21-inch Rolls	Inches of Paper Wasted
Pattern A	0	2	6
Pattern B	1	1	2

2. Answers will vary.

3. Things to consider include the number of 21-inch and 25-inch pages needed, the size of the pages needed for the edition, and the amount of waste produced.

4. The newspaper can control the amount of waste by efficiently choosing the number of large rolls cut using the two patterns.

5. Reducing waste will minimize the cost of producing the newspaper.

6. $w = 6x + 2y$; minimizing

7. No 25-inch rolls; two 21-inch rolls

8. One 25-inch roll; one 21-inch roll

9. Ten 25-inch rolls; twenty 21-inch rolls

10. $y =$ the number of 25-inch rolls; $2x + y =$ the number of 21-inch rolls

11. 25-inch roll requirement: $y \geq 20$; 21-inch roll requirement: $2x + y \geq 50$

12. y; 25-inch rolls are produced using only one of the two patterns.

13. Answers will vary.

14. Answers will vary, but x represents the number of large rolls cut into two 21-inch rolls, and y represents the number of large rolls cut into one 25-inch and one 21-inch roll.

15. Answers will vary, but students should use the objective function $w = 6x + 2y$ to determine the amount of waste.

16. The graphs of the two lines are parallel.

17. The line representing a smaller amount of waste is lower and to the left.

18. (0, 50). It is a corner point (vertex) of the feasible region, as shown here.

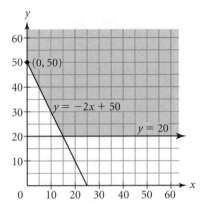

19. No large rolls should be cut into two 21-inch rolls, and 50 large rolls should be cut into one 21-inch roll and one 25-inch roll.

20. $6(0) + 2(50) = 100$. The optimal solution occurs where the first or last parallel line intersects the feasible region. This will always occur at a lattice point, or an edge of the feasible region.

21. The optimal solution is the intersection of the feasible region and the first (or last) line of constant waste that intersects it. For a polygonal feasible region, this intersection must be a vertex or an edge. For a unique solution, it must be a vertex.

22. If they cannot be stored, any extra 25-inch rolls cut would have to be treated as waste.

23. 20; if any extra 25-inch rolls are waste, then it would not be logical to cut more than 20 rolls using pattern y.

24. This produces 20 rolls of each width.

25. 15; 30 additional 21-inch rolls are needed, and each time a large roll is cut using pattern x, two 21-inch rolls are produced.

26. (15, 20); you may want to note that this is another corner point.

27. $6(15) + 2(20) = 130$

Extension 1: All Things Considered

1. $E_{21} = 2x + y - 50$ and $E_{25} = y - 20$

2. The constraints are equalities instead of inequalities.

3. Minimize $w = 6x + 2y + 21E_{21} + 25E_{25}$ subject to $2x + y - E_1 = 50$ and $y - E_2 = 20$.

4. Notice that there are now four variables, and the problem is impossible to solve by geometric methods. However, as long as the problem can be formulated, computer software could be employed to obtain a solution. In the world of industry and government, such software routinely handles problems formulated using tens of thousands of variables acting under thousands of constraints.

Optimal solution: (16667, 10000)

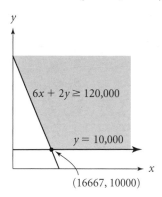

(16667, 10000)

Homework

1. Gauze solutions

a.

5-inch Strips	4-inch Strips	Waste
11	0	0
10	1	1
9	2	2
8	3	3
7	5	0
6	6	1

5-inch Strips	4-inch Strips	Waste
5	7	2
4	8	3
3	10	0
2	11	1
1	12	2
0	13	3

b. Use the first 12 letters of the alphabet, or use subscripts, e.g., $x_1, x_2, \ldots, x_{12}$.

c. Let x = the number of 5-inch strips cut into six 3/4-inch rolls. Let y = the number of 5-inch strips cut into one 3-inch roll and two 3/4-inch rolls. Minimize $w = 0.5x + 0.5y$ subject to the constraints $6x + 2y \geq 120000$ and $y \geq 10000$.

2. Art posters

Sample answers:

	Sheet 1	Sheet 2	Sheet 3
12 × 9	6	6	4
5 × 6	6	6	4
Waste	3 × 12 or 2 × 18	3 × 12 or 2 × 18	12 × 26

	Sheet 1	Sheet 2	Sheet 3
12 × 9	5	5	6
5 × 6	8	8	0
Waste	12 × 4	12 × 4	12 × 18

	Sheet 1	Sheet 2	Sheet 3
12 × 9	8	8	0
5 × 6	0	0	16
Waste	0	0	16 × 24

Many other solutions are possible, and it will be necessary to judge the various rationales of each student's solution.

Activity 4: Trim Loss at the "Cutting Times"

Cutting Patterns

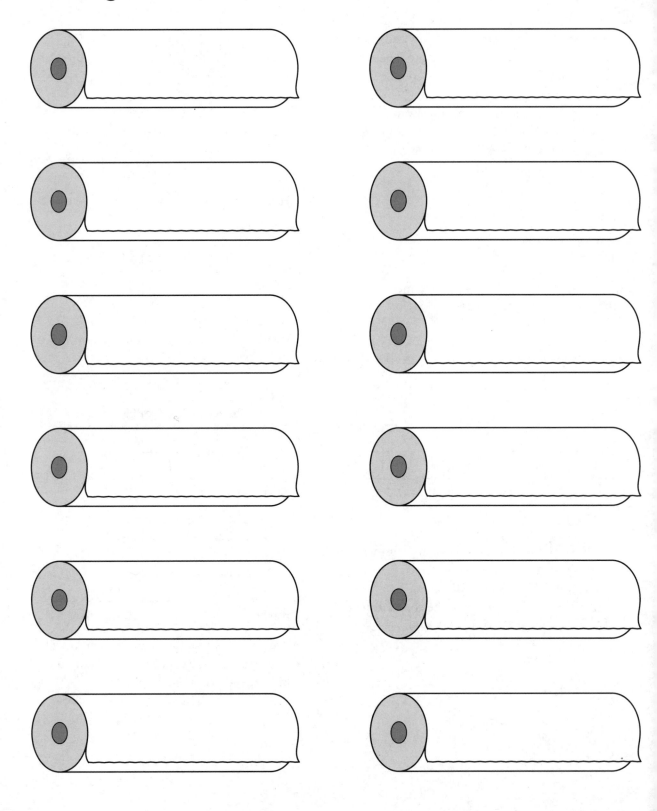

ACTIVITY 5

Workforce Planning at Pizza π

Mathematical programming helps managers find the best ways to optimize resources in order to maximize profits or minimize costs. In the service industry, one important resource is the workforce, or the set of employees. The goal of workforce planning is to have workers with the right skill sets in the right locations when they are needed. Many managers solve these problems using mathematical programming models.

Every mathematical programming model includes a set of decision variables, an objective function to be maximized or minimized, and a system of constraints, or limitations, that restrict the decision maker's options. In some cases, there is also an implied constraint that the answer must be a whole number (for instance, the number of workers must be a positive whole number). To solve these problems, you use *integer programming*.

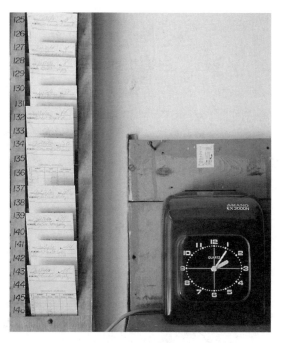

Many hourly employees begin the work day by punching their card in a time clock.

THE WORKFORCE PLANNING SCENARIO

Melanie Johnston is the manager of a Pizza π franchise. In working out her summer staffing needs, she is trying to determine the best mix of experienced and inexperienced workers to hire. Inexperienced workers are paid $6.00 per hour, compared to $8.00 per hour for experienced workers. However, inexperienced workers are only two-thirds as productive as experienced workers, and they can't do some of the important tasks around the restaurant.

Based on many years of experience, Melanie has come to the conclusion that there must be at least one experienced worker for every two inexperienced workers for the operation to run smoothly. Taking into account the lower productivity of inexperienced workers, she estimates that she needs the equivalent of at least ten experienced workers at all times in order to handle the workload during the course of the week. Also, she has made a commitment to the community to hire at least four teenagers from the neighborhood who have no previous work experience.

The Decision Variables

Melanie wonders, What is the optimal workforce that minimizes my total hourly labor costs? To answer this question, she needs to identify the decision variables in the problem.

1. What decisions must she make in this situation?

2. What variable quantities are affected by those decisions?

The Objective Function

Let E represent the number of experienced workers, I represent the number of inexperienced workers, and W represent the total hourly wages.

3. Write an equation that Melanie could use to compute the total hourly wages. This equation is called the objective function, because her objective is to *minimize* the total hourly wages she must pay.

The Constraints

Next, consider the limitations related to Melanie's workforce.

4. In coming up with a plan, what restrictions on the decision variables must Melanie consider?

5. Using I and E, write a mathematical expression that represents her commitment to the community.

6. Using I and E, write a mathematical expression that represents the relationship between the number of experienced and inexperienced workers.

7. If there were nine inexperienced workers, how many experienced workers would there have to be in order to meet the requirement of having the equivalent of at least ten experienced workers?

8. Using I and E, write a mathematical expression that represents the requirement of having the equivalent of at least ten experienced workers.

The Feasible Region

Make a graph with I on the horizontal axis and E on the vertical axis. Graph each of the constraints, and identify the region containing all the points that satisfy all of the constraints. This is the feasible region. Each point in the feasible region might be a solution, because it satisfies all of the constraints. However, only the optimal solution minimizes the average total hourly wages.

9. Determine the coordinates of all of the corner points of the feasible region.

The Optimal Solution

10. In linear programming, the optimal solution must lie at a corner point. In this case, none of the corner points you identified in question #9 can be the optimal solution to Melanie's workforce planning problem. Why not?

11. Use your graph to identify one point in the feasible region that could represent the optimal solution. In terms of the problem situation, what does that point mean?

12. What is the average total hourly wages that Melanie would have to pay if she used the values of I and E corresponding to the point you selected?

13. Now, on your graph, mark each point on or near the boundary of the feasible region that could be the optimal solution, and list its coordinates.

    You could use your list of points from question #13 to search for the optimal solution by finding the total hourly wages associated with each point.

14. If you include the point $(4, 8)$ in your list, you do not need to include the points $(4, 9)$ or $(5, 8)$. Explain why.

    Use this principle to refine the list of points you identified in question #13. Then compute the total hourly wages associated with each point.

15. Which point in the feasible region is the optimal solution?

16. What is the total hourly wages associated with this point?

    To help understand why this point is optimal, set the objective function equal to the optimal total hourly wages and add the graph of that equation to your graph of the feasible region.

17. Use your graph to explain why it makes sense that the point you identified in question #15 produces the lowest total hourly wages. The problem you have solved is an integer programming problem. All of the decision variables must have integer values. Because the optimal solution might not lie at a corner point, as happened here, integer programming problems are more difficult to solve than linear programming problems.

Everyday Applications of Operations Research

Workforce Planners at United Airlines Develop a Scheduling Program to Handle Scheduling for 4000 Employees

Scheduling employees for an airline is an enormous task. There are pilots, flight attendants, grounds people, reservationists, baggage handlers, cooks, janitors, and mechanics. In 1983, United's workforce planners developed a scheduling program for the employees in the reservation offices. The program was used to schedule over 4,000 workers and combined both daily and weekly scheduling in a single model. The model they created consists of 20,000 decision variables and nevertheless can effectively produce monthly shift schedules.

Scheduling flight crews is even more complicated and is considered to be the most difficult scheduling problem in operations research.

➤ For more case studies and background, go to (**www.hsor.org**).

EXTENSION 1: WAGE INCREASES FOR INEXPERIENCED WORKERS

Melanie is concerned that as the minimum wage increases, she may have to pay as much as $6.25 per hour to attract reliable inexperienced workers. Would this increase affect her optimal workforce plan? What if she has to pay as much as $6.50 per hour?

EXTENSION 2: MORE WORKERS NEEDED

Suppose that instead of needing the equivalent of 10 experienced workers at all times, she needs the equivalent of 12 experienced workers. How does this change affect the optimal solution?

EXTENSION 3: MORE SUPERVISION NEEDED

Finally, what if the inexperienced workers needed an even higher level of supervision? Suppose that the ratio of experienced to inexperienced workers has to be one-to-one. How does this change affect the optimal solution?

HOMEWORK

Monday	8
Tuesday	4
Wednesday	6
Thursday	5
Friday	7
Saturday	9
Sunday	8

1. **Nurses' schedules at Bull Run Hospital.** Dr. Nightingale is in charge of the night shift of the emergency room of Bull Run Hospital. Based on historical data, she estimates that she needs these numbers of nurses on duty on each day of the week.

The department currently uses only two five-day shift schedules, Monday through Friday (Schedule A) and Wednesday through Sunday (Schedule B). Obviously, the second schedule is less attractive, so nurses on that schedule are paid 10% more per week. The pay for schedule A is $800 per week and for schedule B is $880 per week.

- **a.** Write a mathematical programming model for this problem.
- **b.** Identify any constraints that are obviously unnecessary for solving this problem.
- **c.** Find the optimal schedule. What is wasteful about this schedule?
- **d.** Dr. Nightingale is considering changing schedule B to Thursday through Monday. What is the new optimal solution?
- **e.** If the hospital can use only two schedules, which pair of work schedules do you think are the best ones possible? (Try to get the most from the overlapping days. Also, assume that any nurse who does not have the entire weekend off will be paid $880 per week.)
- **f.** Now assume that any consecutive five-day work schedule is possible. How many schedules are possible?
- **g.** If a nurse's days off did not have to be consecutive, how many work schedules would be possible?

2. **Linking time periods.** Most workforce planning models look at multiple time periods (weeks, months, quarters, or years) and require equations that link one time period with the next time period. A subscript is used to write a variable. Let

W_t = the number of workers on staff at the beginning of month t

H_t = the number of workers hired during month t

L_t = the number of workers laid off during month t

Q_t = the number of workers who quit during month t

The variables H_t and L_t are under management control, but Q_t is not. The number of workers on staff at the end of a period is a function of all of these.

$$W_{(t+1)} = W_t + H_t - L_t - Q_t$$

In any organization with ranks, such as the armed forces, a separate set of variables is needed for each rank. A new variable would be

P_t = the number of people promoted during the time period

a. Write a similar equation linking the number of sophomores in your school at the beginning of this year with the number of juniors in your school at the beginning of next year. Clearly define each variable.

b. Write a similar equation linking the population of your hometown on January 1 of this year with the population on January 1 of next year. Clearly define each variable.

PROJECT: WORKER SCHEDULING IN YOUR NEIGHBORHOOD

Visit a post office, police department, hospital, bank, or fast-food franchise and interview the person in charge of scheduling workers. Prepare some questions ahead of time to find out what factors they consider and how they do their scheduling. Write a report summarizing your findings. Your report should include:

- a description of how managers decide the number and type of workers for each shift

- a list of all decision variables, both in words and symbols

- a list of all constraints on scheduling, both in words and symbols

- an objective function, in both words and symbols, specifying whether it is to be maximized or minimized

- a description of any difficulties you had in defining constraints or the objective function

Workforce Planning at Pizza π

TEACHER NOTES

Due to the context of the problem, the corner principle that is typically used in the geometric solution of a linear programming problem will not necessarily yield the optimal integer solution. The method we present is a modified trial-and-error solution to an integer programming problem.

OBJECTIVES

- convert word problems into mathematical expressions
- solve systems of linear equations and inequalities, focusing on lattice points
- restrict the domain of a function to integer values

INITIATING THE ACTIVITY

To set up a context and encourage sharing of prior knowledge, allow students to discuss jobs that they hold now or have held in the past. Some possible prompt questions: In these jobs, how were employees paid? (e.g., hourly, by job, salary, etc.) How many hours did employees work in a week? How were employees trained? If you received formal training, did you earn the same hourly rate during training as you did following training? Was a minimum number of workers required during different time periods? Were there any scheduling difficulties? Why?

You might also discuss some real-world applications with students as part of the opening discussion or as part of the closing summary.

➤ For more case studies and background, go to www.hsor.org .

GUIDING THE ACTIVITY

If students have done the other activities in this unit, it should not be difficult to identify the decision variables and the constraints, and to write the objective function. If they have not done the other activities, review or preview these terms before the activity.

The Pizza π problem has only two decision variables, the number of inexperienced and experienced workers to be hired. In this context, these variables are restricted to the integer domain. With only two decision variables, the problem can be graphed in a plane. In more complicated problems involving more than two decision variables, computer programs are needed to find the solutions. In complex-workforce planning problems, there might be many different types of workers, as well as different work shifts. In the Pizza π problem, the goal is to minimize the total hourly wages that a manager must pay her employees. So, using the objective function, we ask students to find the smallest total amount of hourly wages.

In this problem, all of the decision variables must take on integer values. Therefore, the feasible region for this problem includes only the lattice points within the feasible region.

This requirement is more than a mere technicality, because in this case it renders the corner principle inoperable. This is due to the fact that neither of the two corner points has integer coordinates, as required by the context of the problem. In such cases, the optimal solution will lie on or near the boundary of the feasible region. However, the optimal solution to an integer problem is not necessarily the point with integer coordinates that lies closest to the optimal corner point of the feasible region without the integer restriction (see Guiding the Activity for Activity 4). This complexity generally leads to the evaluation of a large number of feasible integer points in order to learn which is optimal.

Students will need to understand that they must find the values of I and E that satisfy all of the constraints, and that I and E must be whole numbers. You may prefer to let students come to this realization on their own when they learn that both corner points involve rational fractions. To ensure uniformity, we have suggested that students graph I on the horizontal axis and E on the vertical axis. You may also want to discuss with the students the effect of graphing these variables the other way around.

The entire activity in which students discover the optimal solution (questions #11–17) could be conducted as a class competition. Students might work individually or in small groups and could be challenged to find the lowest possible value for the total daily wages that meets all of the constraints (i.e., the optimal solution).

Students frequently write, by mistake, that $E \geq 2I$ rather than $2E \geq I$. You may want to suggest that they substitute some numbers for E and I to see if what they have written meets the given condition.

SOLUTIONS

Activity

1. How many inexperienced and experienced workers she must hire

2. The number of inexperienced and experienced workers, total hourly wages, and profit

3. $W = 8E + 6I$

4. Hire at least one experienced worker for every two inexperienced workers; hire at least four inexperienced workers; hire the equivalent of ten experienced workers. (Inexperienced workers are only two-thirds as productive as experienced workers.)

5. $I \geq 4$

6. $2E \geq I$ (Note: Many students will write $E \geq 2I$.)

7. 4; $(2/3)(9) = 6$, and $10 - 6 = 4$.

8. $\frac{2}{3}I + E \geq 10$

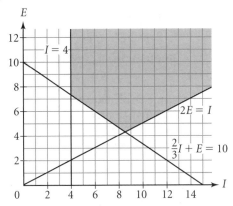

9. $(4, 22/3)$ and $(60/7, 30/7)$

10. Both coordinates must be integers.

11. Answers will vary; e.g., $(4, 8)$, 4 inexperienced and 8 experienced workers.

12. Answers will vary; e.g., $\$6(4) + \$8(8) = \$88$.

13. Answers will vary, but should include $(4, 8)$, $(5, 7)$, $(6, 6)$, and $(8, 5)$.

14. These would produce higher total hourly wages than $(4, 8)$, because $9 > 8$ and $5 > 4$. (Note: There are no values of the form $(7, E)$. The points $(7, 4)$ and $(7, 5)$ are not in the feasible region, and $(7, 6)$ must cost more than $(6, 6)$.)

15. $(6, 6)$. This solution lies on the boundary but, in general, the optimal integer solution need not lie on the boundary.

16. $\$6(6) + \$8(6) = \$84$

Note: Students should have graphed $6I + 8E = 84$.

17. All the other points in the feasible region having integer coordinates lie above the line $6I + 8E = 84$. Those points have higher total hourly wages.

Extension 1: Wage Increases for Inexperienced Workers

Neither increase is large enough to change the optimal solution. (You might challenge students to find an hourly wage for the inexperienced workers

that would affect the optimal solution. Such an hourly wage would have to be greater than $\$8.00$, which does not make sense in the context of this problem.)

Extension 2: More Workers Needed

This change increases the minimum size of the workforce. The constraint now becomes $(2/3)I + E \geq 12$. The new optimal solution is $I = 6$ and $E = 8$, at a cost of $\$100$.

Extension 3: More Supervision Needed

This change has no effect on the original problem, because the optimal solution lies above the graph of the line $E = I$. In general, as the size of the feasible region is reduced by the imposition of tighter constraints, if the optimal solution remains feasible, it will still be the optimal solution.

Homework

1. **Nurses' schedules at Bull Run Hospital**

   a. Let $A =$ the number of nurses on Schedule A. Let $B =$ the number of nurses on Schedule B. Minimize $800A + 880B$ subject to the following constraints listed in order from Monday through Sunday: $A \geq 8$; $A \geq 4$; $A + B \geq 6$; $A + B \geq 5$; $A + B \geq 7$; $B \geq 9$; $B \geq 8$.

   b. The second constraint, $A \geq 4$, is not needed because the first constraint already requires that A be greater than 8. The third and fourth constraints are also redundant because the fifth constraint requires that $A + B$ be greater than 7. The last constraint is also not needed because the next-to-last constraint, Saturday, requires that B be greater than 9. In reality, none of the $A + B$ constraints are needed with this formulation because the minimum value of A or B is greater than the minimum total required.

   c. The optimal solution is $A = 8$ and $B = 9$, for a total cost of $\$14,320$. On Wednesday, Thursday, and Friday there will be 17 nurses on duty, which is far more than the hospital needs.

d. By starting on Thursday and going to Monday, the schedules now overlap on Monday, a busy day. The new optimal solution is $A = 6$ and $B = 9$, for a total cost of \$12,720.

e. The three busiest days are Saturday, Sunday, and Monday. The best pair of schedules would overlap on every one of these three days. Therefore, one schedule should run from Saturday through Wednesday, and the other schedule should run from Thursday through Monday. The Saturday start should have six nurses and the Thursday start should have seven for a total cost of only \$11,440.

f. There are seven different possible schedules, and five sets of nurses could be working on any given day.

g. A schedule consists of selecting the five days to work or, equivalently, selecting the two days not to work. This is an excellent concrete example with which to introduce the concept of combinations. There are $_7C_2$ combinations, for a total of 21 possible work schedules.

2. **Linking time periods**

Sample answers:

a. Let

J_{2006} = the number of juniors at the beginning of next year

S_{2005} = the number of sophomores at the beginning of this year

E_{2005} = the number of new sophomores enrolled this year

L_{2005} = the number of sophomores who left school this year

F_{2005} = the number of sophomores not promoted to juniors at the end of this year

Then $J_{2006} = S_{2005} + E_{2005} - L_{2005} - F_{2005}$.

b. Let

P_{2006} = the population of "our town" on 1/1/2006

P_{2005} = the population of "our town" on 1/1/2005

I_{2005} = the number of people who moved into "our town" during 2005

O_{2005} = the number of people who moved out of "our town" during 2005

B_{2005} = the number of people who were born to residents of "our town" during 2005

D_{2005} = the number of residents of "our town" who died during 2005

Then $P_{2006} = P_{2005} + I_{2005} + B_{2005} - O_{2005} - D_{2005}$.

PROJECT

The project connects the ideas developed in the lesson with methods used by businesses or government in the students' own communities. It can be done individually or in groups over a short period of time.

ACTIVITY 6

Fuel Blending at Jurassic Oil

Peter Yee, the manager of an oil refinery for the Jurassic Oil Corporation, has two grades of gasoline to blend in different ratios. The first grade, purchased from HyOctane, Inc., has an octane number of 92, has a vapor pressure of 4.5 pounds per square inch (psi) at 100°F, and contains 0.4% sulfur. The second grade, purchased from Allif Oil, has an octane number of 85, has a vapor pressure of 5.5 psi at 100°F, and contains 0.25% sulfur. Jurassic Oil wants to produce a blend with these characteristics:

- an octane number of at least 89

- a vapor pressure of no more than 5 psi at 100°F

- a sulfur content of no more than 0.35%

The company wants to produce exactly 120,000 barrels (bbl) of the blend each week.

Grades of gasoline available at the gas pump are a blend of different types of gasoline, each with their own characteristics. A customer selects a grade of gasoline based on price, engine performance, and the car manufacturer's recommendations.

It can purchase no more than 90,000 barrels of each grade per week. Mr. Yee is able to purchase the 92-octane grade from HyOctane, Inc., at a cost of $20 per barrel, and the 85-octane grade from Allif Oil at a cost of $15 per barrel. He must decide how many barrels of each grade to use in the blend to minimize production costs.

In this activity, you will determine the exact quantity of each grade of gasoline that Mr. Yee needs to meet the conditions of the blend, while minimizing the total purchase cost.

Decision Variables

All linear programming problems have decision variables. Use these decision variables to represent the amount of each grade Mr. Yee should purchase.

A = the number of barrels of 85-octane gasoline needed from Allif Oil for the blend Mr. Yee wants.

H = the number of barrels of 92-octane gasoline needed from HyOctane, Inc., for the blend Mr. Yee wants.

1. What decisions must Mr. Yee make?

The Objective Function

In Mr. Yee's gasoline blending problem, he wants to minimize the cost of buying the gasoline needed to produce the blend. That is the goal, or objective.

2. If Mr. Yee purchases 90,000 bbl of gasoline from Allif Oil, how much will it cost?

3. If Mr. Yee purchases 30,000 bbl of gasoline from HyOctane, Inc., how much will it cost?

4. What would be the total cost of the 120,000 bbl?

5. Write an expression to represent the *total* purchase cost, C, for Mr. Yee's blend if he buys A barrels of gasoline from Allif Oil and H barrels of gasoline from HyOctane, Inc.

Constraints

The constraints come from the requirements of the blend. In all, there are six constraints. Two of them deal with the maximum amount of gasoline that Mr. Yee can purchase from the two companies each week.

6. Write two constraints that represent the gasoline purchase restrictions.

7. Write a constraint that represents the *number of barrels* of the blend that Jurassic Oil wants to produce each week. How is this answer different from the other two constraints?

8. Look back at your answer to question #4. What is the *average* cost per barrel of the 120,000 bbl?

9. Explain why the average cost per barrel that you computed in question #8 is *not* $17.50.

In question #8, you found the average cost per barrel. The components of the blend do not contribute equally to the average. So you must use a weighted average formula. The formula for the weighted average cost per barrel is

$$\text{average cost per bbl} = \frac{(A)(\text{cost per bbl of } A) + (H)(\text{cost per bbl of } H)}{A + H}$$

10. Try question #9 again, this time using the formula just given, with $A = 90,000$ and $H = 30,000$. Did you find the same average cost per barrel?

Now we will use weighted averages to express the constraints.

11. The vapor pressure of the blended gasoline must be no more than how many psi? This is a combination of the vapor pressure of the gasoline purchased from Allif Oil Company and of the gasoline purchased from HyOctane, Inc. What vapor pressure does each gasoline have?

12. Use the weighted average formula to represent this constraint.

13. Use the same approach to determine the constraints related to sulfur content and octane rating.

14. For ease in graphing, rewrite these constraints in a standard linear form.

The Feasible Region

The feasible region for Mr. Yee's blending problem is the set of points bounded by the constraints. The optimal solution is within this region. We will find the feasible region for Mr. Yee's blending problem by graphing the six constraints on the same coordinate axes. For consistency with your classmates, let the horizontal axis represent the number of barrels of gasoline purchased from Allif Oil (A), and let the vertical axis represent the number of barrels of gasoline purchased from HyOctane, Inc. (H). On a sheet of graph paper, graph each of the six constraints. Indicate the feasible region on your graph by shading it.

15. What is the shape of the feasible region?

16. Which constraints determine the feasible region?

Redundant Constraints

Sometimes a constraint has no effect on the feasible region. These constraints can be removed from the system without changing the feasible region. They are called **redundant constraints.**

17. Are any of the constraints in Mr. Yee's blending problem redundant constraints? If so, which ones?

18. Is (50000, 70000) a point in the feasible region? Why or why not?

19. Find the endpoints (to the nearest tenth of a barrel) of the feasible region.

20. Use the objective function to find the cost for the point (50000, 70000).

21. Choose another point in the feasible region. Use the objective function to find the cost for this second point.

22. Now test the two endpoints that you found in question #19. Find the cost for each point. Which of the four points you tested yields the lowest cost?

The Optimal Solution

To find the optimal solution, use the graph of the objective function. Set the objective function equal to the cost you found in question #20. Add the graph of this equation to your graph of the feasible region, and label it line *l*. Now, using your answer from question #22, set the expression that defines the objective function equal to the *lowest cost*, graph the equation, and label it line *m*.

23. What do you notice about lines *l* and *m*?

   The *optimal line, m,* intersects the feasible region and has the lowest possible cost for the gasoline. Notice that any line parallel to the optimal line that has a lower cost fails to intersect the feasible region. Any other line parallel to the optimal line that intersects the feasible region has a higher cost. The optimal solution occurs at the point where the optimal line intersects the feasible region.

24. What are the coordinates of the optimal solution? How would you describe the location of this point in the feasible region? What does this point mean in terms of the problem?

25. What are the octane number, vapor pressure, and sulfur content of the optimal blend?

26. What is the lowest amount that Jurassic Oil can pay for its oil?

Everyday Applications of Operations Research

Texaco Uses OMEGA System, Profits Up 30%

During the 1980s, Texaco Oil Company announced a 30% increase in profits—a gain of $30 million. Sources at the petroleum giant said the increase was due to the use of OMEGA, a nonlinear programming system for blending gasoline. The system, which involves 40 variables and 71 constraints, is currently in place at seven Texaco refineries.

➤ For more case studies and background, go to [www.hsor.org].

EXTENSION 1: VAPOR PRESSURE

HyOctane and Allif's products have changed. All the conditions are the same except for the vapor pressure. The grade from HyOctane now has a vapor pressure of 5.5 psi. The grade from Allif has a vapor pressure of 4.5 psi. What is Jurassic Oil's optimal blend now?

EXTENSION 2: CHANGING THE MAXIMUM SULFUR CONTENT OF THE BLEND

Jurassic Oil has an interest in reducing the sulfur content of the gasoline it sells. Mr. Yee was asked to determine the impact of lowering the maximum sulfur content of the blend to 0.34% sulfur. Modify the original problem to reflect this tighter constraint, and solve the new problem.

HOMEWORK

1. **Milkfat blends.** Each week, the DeeLite Milk Company gets milk from two dairies and then blends the milk to get the desired amount of butterfat for the company's premier product. Dairy A can supply at most 700 tons of milk averaging 3.7% butterfat and costing $240 per ton. Dairy B can supply milk averaging 3.2% butterfat and costing $200 per ton. How much milk from each supplier should DeeLite use to get 1000 tons of milk with at least 3.5% butterfat at minimum cost? What is the total cost?

2. **Healthy plant blends.** A fruit grower can purchase two brands of fertilizer. Brand A contains 16% nitrate, 20% phosphate, and 12% sulfate, by weight. Brand B contains 11% nitrate, 50% phosphate, and 2% sulfate. Brand A costs $10 per pound and brand B costs $8.50 per pound. The grower wants to make a mixture of these two fertilizers with at least 14% nitrate, at least 30% phosphate, and at least 6% sulfate.

   How many pounds of each brand of fertilizer should the grower purchase to make 100 pounds of the mixture at minimum cost? What is the total cost?

PROJECT 1: MENU PLANNING WITH CONSTRAINTS

Interview the food service manager at your school and research the recommended daily nutritional allowances for a teenager and the prices of supplies. Design healthy lunch menus for one week for your school cafeteria.

PROJECT 2: PRODUCT MIX AT A NURSERY

Visit a greenhouse or nursery and learn how they determine how many vegetables, plants, shrubs, trees, and so on to grow. Also find out how they determine the appropriate fertilizer mixes for each of the different types of plants. Write a report describing what you learned.

ACTIVITY 6

Fuel Blending at Jurassic Oil

Students will use linear programming to solve a "blending" problem. This problem occurs whenever a company wants to manufacture a product with specific, unique characteristics, unlike any that is available, but can produce such a product by combining two or more products that are available.

OBJECTIVES

- use weighted averages
- consider a feasible "region" that lies wholly on one straight line (the demand equality)
- experience the interdisciplinary nature of problem solving in industry (For example, an understanding of chemistry, materials science, environmental science, and nutrition may be needed in setting up an appropriate system of constraints.)

INITIATING THE ACTIVITY

Allow students to research some of these questions ahead of time, either individually or in groups.

- How do sausage makers mix the various fillings (chicken, beef, pork, fat, spices) to produce a hot dog that is low in cost yet flavorful?
- How is gold mixed with various other metals to produce the different karat weights needed in designing and producing jewelry?
- How do you decide how much of each snack item to place in a mix in order to produce a party mix that is low in cost yet meets public demand?
- How are different blends of gasoline created?
- How do chocolate bar manufacturers minimize expensive ingredients, such as cocoa, and maximize volume, chewiness, sweetness, and other qualities that consumers look for? (They use fillers and emulsifiers, and they blend and sometimes whip the chocolate.)

GUIDING THE ACTIVITY

For uniformity, we have suggested that students graph A on the horizontal axis and H on the vertical axis. You may wish to discuss with students the effect of graphing these variables the other way around, or ask several students to do so.

In these problems it *appears* that A and H must be whole numbers, because fractional parts of a barrel are apparently not sold. However, a barrel is a unit of measure. The amount of gasoline purchased may be delivered by a large tanker, or even via a pipeline, so that fractional barrels *could* be purchased. Thus, this is truly a linear programming problem and not an integer programming problem. For graphing the constraints, students may find it helpful to express all of the constraints in slope-intercept form—that is, $H = mA + b$. Students may solve for the corner points either algebraically or by using the "calculate intersection point" feature of a graphing calculator.

Once students have conceptualized the problem situation, the entire activity could be conducted as either a cooperative learning situation or a competition. Groups might represent competing oil companies, each working at finding the lowest-cost blend that meets all of the constraints.

Students are asked throughout to express answers in standard linear form. This matches the form operations researchers customarily use. Although we have treated octane number as a linear function, it is only approximately linear in this case. This nonlinearity is explored in one of the extensions available on the website.

Question #9 sets up the concept of a weighted average, which is used next, so you may want to be sure that all of your students understand why the average cost per barrel of the gasoline purchased for the blend is less than the average of the two costs per barrel. Also, point out that the denominator is $A + H$, not 120,000, because that constraint may change.

Students are asked to graph the six constraints. This is possible because there are only two decision variables. In situations involving more decision variables, it is necessary to use a computer to determine the solution.

After graphing the six constraints, students should notice that the feasible region is a line segment. Most linear programming problems result in feasible regions that are polygons. Because one of the six constraints in this case is an equation, not an inequality, the "region" is reduced to a line segment.

Two of the six constraints, $A \leq 90,000$ and $H \leq 90,000$, are not boundaries of the feasible region and are therefore redundant. A discussion topic could be, For what values of n would $A \leq n$ and $H \leq n$ become boundaries of the feasible region? This is a simplified example of what operations researchers call **sensitivity analysis**. Sensitivity analysis refers to a process of assessing the degree to which the optimal solution is sensitive to changes in key model parameters.

EXTENSIONS

The first extension demonstrates the possibility of there being no feasible region.

➤ For more extensions with solutions, go to [**www.hsor.org**].

SOLUTIONS

Activity

1. He must decide how much of each type of gasoline to purchase from each company.

2. $(\$15)(90{,}000) = \$1{,}350{,}000$

3. $(\$20)(30{,}000) = \$600{,}000$

4. $\$1{,}950{,}000$

5. $C = 15A + 20H$

6. $A \leq 90{,}000$; $H \leq 90{,}000$

7. $A + H = 120{,}000$. It is an equation, not an inequality.

8. $\$1{,}950{,}000/120{,}000$ barrels $= \$16.25$ per barrel

9. More gasoline was purchased from Allif than from HyOctane, so the average cost per barrel should be closer to the selling price of HyOctane, not the average of the two selling prices.

10. Yes, the answer is the same.

11. 5; Allif is 5.5; HyOctane is 4.5.

12. Combined vapor pressure $= \frac{5.5A + 4.5H}{A + H} \leq 5$.

13. Sulfur: $\frac{0.25A + 0.4H}{A + H} \leq 0.35$; octane: $\frac{85A + 92H}{A + H} \geq 89$.

14. Vapor: $0.5H - 0.5A \geq 0$, or $A - H \leq 0$; sulfur: $0.05H - 0.1A \leq 0$, or $2A - H \geq 0$; octane: $3H - 4A \geq 0$, or $4A - 3H \leq 0$.

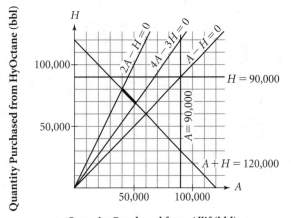

Quantity Purchased from Allif (bbl)

15. A line segment

16. $2A - H \geq 0$ $(H \leq 2A)$;
    $4A - 3H \leq 0$ $(H \geq \frac{4}{3}A)$;
    $A + H = 120{,}000$ $(H = 120{,}000 - A)$

17. $A \leq 90{,}000$; $H \leq 90{,}000$; $A - H \leq 0$ $(H \geq A)$

18. Yes, it meets all of the constraints.

19. $(51428.6, 68571.4)$ and $(40000, 80000)$

20. $\$15(50{,}000) + \$20(70{,}000) = \$2{,}150{,}000$

21. Answers will vary, but to be feasible, the point must be on the feasible line segment.

22. $(51428.6, 68571.4)$: $C = \$15(51{,}428.6) + \$20(68{,}571.4) = \$2{,}142{,}857$

    $(40000, 80000)$: $C = \$15(40{,}000) + \$20(80{,}000) = \$2{,}200{,}000$

    $\therefore (51428.6, 68571.4)$

23. The lines are parallel.

24. $(51428.6, 68571.4)$. It is an endpoint of the feasible line segment. Purchasing 51,428.6 barrels from Allif Oil and 68,571.4 barrels from HyOctane will yield the minimum cost.

25. Octane number: 88.999998, which is approximately 89; vapor pressure: 4.929 psi; sulfur content: 0.3357.

26. $\$2{,}142{,}857$

Extension 1: Vapor Pressure

No solution is possible. The "new" vapor pressure constraint is $H - A \leq 0$. It is impossible to meet all of the constraints. When the feasible region is the empty set, there does not exist an optimal solution to the proposed situation. You may wish to have the students sketch the graphs and discuss what could be done in this case.

Extension 2: Changing the Maximum Sulfur Content of the Blend

This change replaces the constraint $H \leq 2A$ with $H \leq 3A/2$, which reduces the size of the feasible region. However, the endpoint that represents the optimal solution is unaffected. Therefore, this change has no impact on the optimal solution.

Homework

1. Milkfat blends

Let A = the number of tons purchased from dairy A, and let B = the number of tons purchased from dairy B. The problem is then to minimize $C = 240A + 200B$ subject to $A \leq 700$, $A + B = 1000$, and $3B - 2A \leq 0$. The feasible region is a line segment with endpoints (600, 400) and (700, 300). Most students will probably see that 600($240) + 400($200) < 700($240) + 300($200), so that (600, 400) is optimal. The total cost of the blend is $224,000.

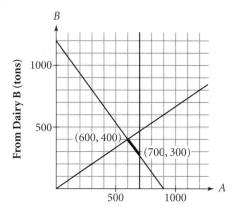

From Dairy A (tons)

2. Healthy plant blends

Let x = the number of pounds of brand A used in the mixture, and let y = the number of pounds of brand B used in the mixture. The problem can then be formulated:

Minimize the objective function, $C = 10x + 8.5y$, subject to the constraints $x + y = 100$, $y \leq \frac{2}{3}x$ (nitrate content), $y \geq \frac{1}{2}x$ (phosphate content), and $y \leq \frac{3}{2}x$ (sulfate content).

Note that the sulfate constraint is redundant.

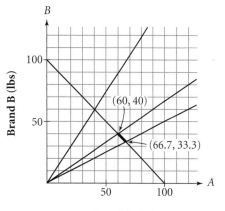

Brand A (lbs)

The feasible region is a line segment with endpoints (60, 40) and (66.$\bar{6}$, 33.$\bar{3}$). Next, the minimum cost must occur at one of these endpoints. Computing the cost for each, $C = 10(60) + 8.5(40) = \$940$, and $C = 10(66.\bar{6}) + 8.5(33.\bar{3}) = \$949.99$, we find that the optimal solution is to purchase 60 pounds of brand A and 40 pounds of brand B. Students may see that it is unnecessary to compute the cost associated with each endpoint, because using less of the more costly brand A and more of the less costly brand B will clearly reduce cost.

Queuing Theory

BACKGROUND

All of us have experienced the annoyance of waiting in line. We wait in line in our cars in traffic jams or at tollbooths; we wait on hold for an operator to pick up our telephone calls; we wait in line at supermarkets to check out; we wait in line at fast-food restaurants; and we wait in line at banks and post offices. As customers, we generally do not like these waits. Managers also do not like us to wait, because it may cost them business. Why, then, is it necessary for us to wait?

The answer is relatively simple: There is more demand for service than there is ability to provide service. There may be a shortage of available servers or a limited number of hours of service that can be provided. However, even with a seemingly sufficient and capable service staff, delays are sometimes unavoidable. For instance, suppose it takes 30 minutes to serve a customer, and two customers arrive within the same hour. A quick calculation makes it seem that both can be easily served within an hour. However, if they arrive close together, one will have to wait almost half an hour!

To know how much service should be made available, you need to take randomness into account. You also need to know the answers to such questions as, How long does a customer wait? and How many people are in the line? **Queuing theory** uses this information and detailed mathematical analysis to help determine the appropriate level of service needed.

The word "queue" is used in Great Britain and elsewhere to mean "waiting line." The term "queuing theory" is used by engineers and mathematicians to refer to the study of service lines and wait times. Regardless of whether you call it a queue or a line, waiting in one can be unpleasant!

A **queuing system** can be described simply as customers arriving for service, waiting for service in a line (or queue), and leaving the system after being served. The term "customer" is used in a general sense and does not necessarily imply a human customer. For example, a customer could be a ball bearing waiting to be polished, an airplane waiting to take off, a computer program waiting to be run, or a homework assignment waiting to be done.

A NOTE ABOUT THE PSYCHOLOGY OF QUEUING

Many companies (Disney theme parks are one example) have become expert in understanding the psychology of waiting. Waiting in a line that is moving seems less boring than standing still in the same spot. Television monitors can help keep visitors' minds off the long wait. In addition, if they can see and hear some of the excitement of those who have completed their wait, anticipation increases and waiting seems worthwhile. Expectations are another major factor in customer satisfaction. If customers are told ahead of time that the wait will be 15 minutes, at least they can make an informed decision to join the line or not. If it turns out to be less than the quoted 15 minutes, they are pleasantly surprised.

Another dimension to the psychology of waiting relates to fairness. When there are several lines, you might get stuck behind a customer who has a complicated request that takes a long time to service. As a result, someone waiting in a different line might end up waiting less time. Many organizations have addressed this potential inequity by creating one line that all arriving customers enter. Thus, anyone who arrives after you must be farther back in line and cannot be served before you.

ASPECTS OF A QUEUING SYSTEM

The key features of queuing systems are characteristics of arrivals, service discipline, and service characteristics.

Characteristics of arrivals. Usually, the timing of arrivals is described by specifying the average rate of arrivals, *a,* per unit of time, or the average inter-arrival time, $\frac{1}{a}$. For example, if the average rate of arrivals, *a,* is 10 per hour, then the inter-arrival time, on average, is $\frac{1}{a} = \frac{1}{10}$ hour = 6 minutes.

In many simple applications, inter-arrival times, *t,* have an exponential distribution of the form $f(t) = ae^{-at}$. Other arrival patterns are not exponential, such as a doctor's scheduled appointments. In such cases, the formulas given here do not apply and other queuing models must be used.

Queue Discipline and Structure. The queue discipline is the rule, or set of rules, specifying which of the waiting customers is next to receive service. The most common queue discipline is first-come-first-served. Other queue disciplines include last-come-first-served (inboxes), service-in-random-order (retail stores), and shortest-processing-time (express lines).

A system may include one or more servers. However, when there is more than one server, waiting customers may either form a line before each server or form one common waiting line.

Service characteristics. Usually, each customer is served by one and only one server, no matter how many servers there are. The service time is assumed to be random and exponentially distributed, and when there are multiple servers, it is assumed that all servers are identical. That is, we assume that each is able to service customers at the same average rate, *h.* (It might seem more natural to use *s* or perhaps *c* to represent this variable, but *s* and *c* are used for other variables in queuing theory. Therefore, throughout Activity 7, we have used the term "help" in place of "serve.") The reciprocal, 1/*h*, of the average rate of service is the average time required to serve one customer. For example, if a server can serve 3 customers per hour on average, then *h* = 3 and 1/*h* = 1/3 hour. So 20 minutes is the average service time for one customer.

QUEUING FORMULAS

Analysis of queues requires defining certain performance measures. Each is an average.

L = average length of the line

W = average waiting time

a = average arrival rate

h = average help rate (the rate at which a customer can be helped)

$1/a$ = average time between successive arrivals

$1/h$ = average service time per customer

x = average number of arrivals during the average service time (also called **traffic intensity**)

The traffic intensity is the ratio of arrival rate to help rate, or $x = a/h$. This ratio also gives the average number of arrivals during an average service time. So, for example, if the average arrival rate a = 10/hour and the average service time $1/h$ = 1/2 hour, then x = 5 represents the average number of customer arrivals during the average service time of 30 minutes. In this case, the traffic intensity is high: Since $x > 1$, the line will grow indefinitely. For $x < 1$, this ratio also represents the fraction of time the server is busy; for $x \geq 1$, the server will be busy 100% of the time.

Queuing theorists have discovered that in a single-server queuing system in which customers arrive at random and service time is exponentially distributed, the long-term average number of customers in the system, L, is given by the formula

$$L = x/(1 - x) = a/(h - a)$$

There is also a direct relationship between the length of the line, L, and the waiting time, W. The formula is known as Little's law: $L = aW$ or, equivalently, $W = L/a$.

CASE STUDY: VIRGINIA DEPARTMENT OF MOTOR VEHICLES

In 1995, the Virginia Department of Motor Vehicles (DMV), disturbed by numerous customer complaints at all facilities, instituted a plan to decrease customer wait time.

A new system was designed to have customers form one or two lines upon entering the building. The servers for the lines assess the customer's request for service and assign a ticket on which each job is letter-coded by type. Also on this ticket is a number representing the order in which the customer will be called for service within this job category. Customers are provided with a clipboard, all the paperwork needed for their job, and a seating area. This arrangement contributes to a more effective service arrangement and a relaxed climate for customers.

In February 1997, in the Arlington, VA, DMV facility, 17,929 customers were served, with an average wait time of 21 minutes. After the implementation of the new system in Arlington in February 1998, 20,843 customers were served, with an average wait time of 10 minutes, 12 seconds. A significant by-product of the new system was a decrease in employee stress.

References

Smoot, Janet, Bill Jacobs, and Michelle Fadely. 1997. *Virginia Department of Motor Vehicles queuing management system.* Richmond, VA: DMV.

➤ To learn more about the Virginia DMV and other case studies, go to ⎡ **www.hsor.org** ⎤.

ACTIVITY 7

Does This Line Ever Move?

Jasmine Billings is vice president in charge of operations for Arm-and-a-Leg Ticket Sales. She is concerned about complaints regarding long waits at the ticket windows on Friday afternoons at many of the malls. To reduce the number of complaints, Jasmine Billings has hired Dr. Kyle Ross, an expert on queuing theory.

Queuing theory deals with the mathematical study of waiting lines. In analyzing this problem, Dr. Ross makes these assumptions:

- individual customers arrive at random to purchase tickets

- the time to complete a purchase is also random. This might be due to the number of tickets the customer purchases or the customer asking for information about dates and seat location.

Collecting Data

Dr. Ross spent several Friday afternoons observing the situation and collecting the data. He found that the average number of customers arriving per hour is 18 and that the average number of customers a single ticket agent can help per hour is 20.

1. What variables did Dr. Ross need to observe during his Friday afternoon visits to the mall?

2. If 18 customers arrive per hour, on average, how much time elapses between successive arrivals?

3. If a customers arrive per hour, on average, how much time elapses between arrivals?

4. If 20 customers are helped per hour, on average, how long does it take to help one customer?

5. If h customers are helped per hour, on average, how long does it take to help one customer?

Modeling the Queue

To develop a mathematical model of our queue, we must have an idea of the traffic intensity, x, which is the ratio of the average rate of customer arrivals, a, to the average rate of customers being helped, h. In order for this ratio to make sense, the time units of a must be the same as those of h.

6. Write a ratio between the variables that represents the traffic intensity. Use the values of these variables to find the value of x.

7. The average number of customers in the system, L (including those in line and the one at the ticket window), can be represented by the function

$$L = \frac{x}{1 - x}$$

Using the value of x from question #6, calculate L. What does this value of L tell you about this queuing system?

Customers are usually more concerned with the length of time it takes to get a ticket than the length of the line. Therefore, let W represent the average time a customer waits in the system, including the time in line and the time to be helped. The function $L = aW$ expresses the relationship between L and W.

8. What are the units for each of the variables a, L, and W?

9. Given that $L = aW$, use your values of a and L to find the value of W for the system at Arm-and-a-Leg Ticket Sales. What does this value of W tell you?

Using the Model

Suppose Dr. Ross learns that during a slow time of the day an average of only 16 customers arrive per hour to purchase tickets. However, the customer help rate remains the same.

10. What is the traffic intensity, x, during this time of day? Use this new value of x and calculate the values of L and W for this time of day.

11. Record the values you have so far for x, L, and W in the appropriate places in the following table. Then complete the rest of the table for the given values of a and h.

a (customers/hour)	h (customers/hour)	x	L (customers)	W (hours)
14	20			
16	20			
18	20			
18	22			
18	24			

12. Using the values you entered in the table in question #11, describe what happened to the values of L and W when a increased while h remained constant.

13. What happened to the values of L and W when h increased while a remained constant?

14. If $a = 22$ and $h = 20$, what are the values for x, L, and W? Do all of these values make sense? Explain.

15. Why did this particular value of x occur? Explain what would really happen to the ticket line, given the situation in question #14.

16. Compare the value of x in question #14 with all of the values of x in the table. What do you observe?

Interpreting the Graph

The graph of $L = x/(1 - x)$ is shown at right.

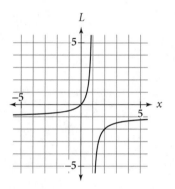

17. Using the equation for L, describe what happens when $x = 1$. Where is this represented on the graph?

18. For the function $L = x/(1 - x)$, what is the domain of x?

19. If $L = -1$, then $-1 = x/(1 - x)$. What happens when you solve this equation for x? Where is this represented on the graph?

20. For the function $L = x/(1 - x)$, what is the range of L? What happens to the value of L when $x > 1$? Discuss why these values of L make sense, or do not make sense, for the Arm-and-a-Leg problem.

21. Thinking about the definition of x as the ratio a/h, what would a value of $x = 0$ mean? What would a value of $x < 0$ mean?

22. Trace the portion of the graph shown earlier that is appropriate for the Arm-and-a-Leg problem. Write the domain for the portion of the graph you traced.

23. Calculate and enter the value of L for each of the values of x in a table like the one shown. What happens to L as the value of x approaches (gets closer to) $x = 1$? Where is this represented on the graph?

x	L
0.8	
0.9	
0.95	
0.99	

Applying the Model

24. In all of the analyses above, changes were made in the values of a and h. In reality, which, if any, of these variables would a manager like Jasmine Billings be able to control?

25. What could Jasmine Billings do to improve customer satisfaction at her sales outlets? What strategies for improving customer satisfaction have you experienced while waiting in line?

Everyday Applications of Operations Research

L. L. Bean Improves Service and Minimizes Lost Calls

L. L. Bean is a retailer of outdoor goods and apparel. Sixty-five percent of its goods are sold through phone orders, and the three weeks prior to Christmas make or break the year financially.

In 1988, management learned that, on average, 80% of callers received a busy signal when they used the toll-free number to place an order, and those customers who got through were on hold for an average of 10 minutes. L. L. Bean's long-distance charges for its toll-free line often amounted to $25,000 per day. The loss in sales from customers who hung up were unknown.

L. L. Bean hired a consulting team that used queuing theory to develop a model focused on improving the efficiency of its telemarketing operations. The model determined optimal levels for the number of phone lines carrying incoming calls to telephone agents, the number of agents scheduled, and the queue capacity, or the number of calls that can be waiting.

The queuing analysis process cost L. L. Bean $40,000, but it estimated an increased profit of $10 million in 1989. This model improved service rates and call volume throughout the year, especially during the peak period just before Christmas.

➤ For more case studies and background, go to [**www.hsor.org**].

HOMEWORK

1. **The queue at the W.C.** At an outdoor concert, there are two portable toilets for women (one at each end of the field), and one portable toilet for men. Assume that the average customer arrival rate at the men's toilet is 30 per hour, and the average customer arrival rate at each of the women's toilets is 15 per hour. Assume also that the average service time for men is 1 minute and for women is 3 minutes. Are the portable toilets distributed fairly? Use queuing theory to support your answer.

2. **The book checkout.** At a local library, one clerk is checking out books. On average, 40 people per hour arrive at the counter to check out books. It takes an average of 1 minute to help a person.

   a. On average, how many people are in the queuing system?
   b. On average, how much time will a person spend in the queuing system?
   c. On average, how much time will a person spend in line before reaching the counter?

PROJECT 1: THE PSYCHOLOGY OF WAITING

Write a two-page report on the psychology applied to a recent experience you had waiting in line for a long time. Describe the queuing situation, the layout of the service facility, the number of servers, and the organization of the queue. Describe your wait: Was it longer than expected? Did you find the wait boring? What did you do to pass the time while waiting? If you had known how long the wait would be, would you have done anything differently? Describe management's approach to the queuing situation: Did the management do anything to make the wait more pleasant? Was there anything else management might have done to reduce the average waiting time or improve your experience?

PROJECT 2: QUEUING SYSTEMS AT WORK

Make a list of questions and visit a service facility in your community. Make an appointment to interview the manager about any problems that have been encountered with respect to how long customers have to wait and the customers' satisfaction. Inquire about solutions that have been tried and how effective each solution has been. Some suggestions of places to visit are a post office, bank, grocery store, school cafeteria, theater, fast-food franchise, or restaurant.

PROJECT 3: FIELD RESEARCH

This project requires a group of four. Each group should select a different business in the community. Each member of the group should be assigned an interval of 1 to 3 hours over which to observe the business and gather data on how often customers arrive and the length of time it takes for each customer to be served. Then compile the group data into a single table using the variables in queuing theory. Write a report noting your observations and whether you think the business is using queuing theory. If it is, what queuing model is being used? Make a diagram to illustrate the queuing system observed.

ACTIVITY 7

Does This Line Ever Move?

The techniques of queuing theory are used to solve problems in business, industry, and government. Many of these contexts will be familiar to students.

OBJECTIVES

- model to simulate real-world problems
- advance the understanding of randomness and its effects
- work with rational functions

INITIATING THE ACTIVITY

You may want to open this activity with a discussion of students' personal experiences in a variety of queuing situations. Some questions you might ask are

- What was the longest you ever waited in line, and why was there a long wait?
- What would have improved the wait?
- Did you ever spend a long time waiting in line without minding the wait?
- Describe examples of waiting for service from a nonhuman server.

   Many aspects of queuing theory are familiar to students; however, some may not be familiar or intuitive. For example, students may think that if $h > a$, there will be no waiting, but an arrival rate of 18 and a help rate of 20 can mean average wait times of 30 minutes! That is because actual arrival rates can vary greatly from the average at certain times. Also, customers rarely arrive at even intervals. Because the queuing model used in this activity assumes that both customer arrivals and customer service times are random, understanding the meaning of randomness is crucial to understanding this queuing model. Furthermore, students may better appreciate the key concepts in a queuing model if they first simulate a queuing situation. We strongly suggest using these pre-activities before doing Activity 7. In using Pre-activity 2, you may prefer to have students run through the simulation once without measuring anything, and then follow up with a discussion about what quantities it would make sense to measure.

TEACHER NOTES

Does This Line Ever Move?

83

PRE-ACTIVITY 1: COIN FLIPS *(10 minutes)*

You will need a coin or two-color counter for each student. The purpose of this activity is to compare human notions of randomness with the actual results of 20 coin flips.

1. Students simulate flipping a coin 20 times by pretending to do 20 coin flips and writing a "result," H or T. Label this sample List A.

2. Next, students *actually* flip a coin 20 times and record H or T for each trial. Label this sample List B.

3. Place students in groups of four. Give each group four simulated and four actual samples. Compare and contrast the two different samples. Describe the patterns in each set. Emphasize the presence of longer strings of repetitive data (e.g., four or more heads in succession) when the coins are actually flipped. (Long strings are generally absent when humans attempt to simulate random results.)

4. Suppose an H represents a customer arriving for service. How likely is it that customers will arrive at regular intervals? (Very unlikely. There will be long lulls, and frequently several people will arrive, one after the other.)

PRE-ACTIVITY 2: QUEUE AT THE SHARPENER *(15 minutes)*

You will need a clock with a second hand (or three stopwatches), calculators with random number generators (or copies of the random integer tables), a pencil sharpener, and pencils. With a small class (15 or fewer), a queue might not form. In that case, you may want to have students sharpen (or simulate sharpening) two pencils, rather than just one.

➤ More random integer tables are available online at (**www.hsor.org**).

Simulate a queue at the pencil sharpener to illustrate vocabulary used in queuing theory: average number of customers in a system and average wait time.

1. Select three student recorders. Each remaining student generates a random number from 1 to 150 to determine when he or she will enter the line to use the pencil sharpener. At the indicated time, in seconds (the random integer), the student should be ready to enter the queue and sharpen (or simulate sharpening) a pencil.

2. Run the simulation for 2 minutes.

3. The first recorder collects data to determine the average number of people in the line. This person records the number of people at the pencil sharpener (including the person sharpening) at 30, 60, 90, and 120 seconds.

T E A C H E R   N O T E S

The average of these four numbers approximates the average number of people in the system at any given time.

4. The second recorder collects data to determine the average time in the system. This person records the wait time for every fifth person to join the line and calculates the average of these wait times to approximate the average wait time. The time in the system includes the time in line and the time used to sharpen the pencil. (Note: You may need a "judge" at the pencil sharpener who decrees when a pencil is sharp enough. Or you can let students who take too long represent "difficult customers.")

5. The third recorder collects data to approximate the average service time. This person records the time it takes each fifth person to sharpen (or simulate sharpening) a pencil and computes the average of these times.

6. After the simulation, calculate the average arrival rate: Count the number of people who got to stand in line and/or use the sharpener (by a show of hands), and divide by 2 minutes. Also ask students to express the average arrival rate in customers per hour. Discuss any changes as the situation approached a steady state.

7. If you have time, repeat this simulation with different sets of random numbers. Otherwise, speculate what would happen if the simulation were repeated.

8. Average the data for each trial: the average number of people in line, the average time in the system, and the average arrival rate.

RANDOM INTEGER TABLES (1–150) FOR TRIAL #1

22	145	18	50	48	80	6
46	3	124	40	127	24	61
56	85	42	36	63	49	140
56	62	47	9	18	21	40
25	45	145	23	50	75	120

RANDOM INTEGER TABLES (1–150) FOR TRIAL #2

122	27	109	68	94	46	140
16	75	128	44	58	126	111
4	24	36	125	14	124	91
3	91	135	127	47	12	81
140	48	10	25	99	5	93

GUIDING THE ACTIVITY

Remind students of the initiating activities to reinforce the nature of randomness and its effect on the pattern of customer arrival times and service times. You may want to contrast these random patterns with a hypothetical case where the arrivals were evenly spaced and the service times uniform.

Question #1 should be carefully guided. You may want to ask the students what actual data Dr. Ross might have collected to arrive at the average rates of 18 and 20. (Answers will vary, and they may or may not include "18 customers arrive during any one-hour period of time.")

Be sure *everyone* defines the variables in question #1 as

$a = 18$ customers arriving per hour

$h = 20$ customers helped per hour

so that the variable names are consistent for the rest of this activity.

Reinforce the correct labeling of units, and remind students that their time units must be consistent.

In question #3, if you want the average interval between customer arrivals, then using the units as a guide, hours per arrival is the reciprocal of arrivals per hour.

In question #7, you may want to leave the answer as a common fraction to emphasize the simplification of complex fractions that will be encountered later.

Note that in question #9, the units of L will be customers, even though the formula $L = x/(1 - x)$ yields a unitless answer.

In questions #13 and #14, you can explore the effect of a small change in a on the values of L and W.

For question #14, students can fill in the values of x, L, and W at the overhead. The table can also be used to prompt answers for questions #15 and #16. Encourage class discussions by continually referring to what would be happening at the ticket sales office.

Questions #18 through #23 refer to the function $L = x/(1 - x)$ in the real number system, including $x < 0$, and not just to the limits defined by queuing theory. In question #17, you may want to introduce the concept of a vertical asymptote. Likewise, in question #19, you could introduce the concept of a horizontal asymptote. In question #18, as in question #17, the function $L = x/(1 - x)$ is not defined at $x = 1$ and, therefore, the domain is restricted to all real numbers except $x = 1$.

SOLUTIONS

Activity

1. The number of customers arriving and the number of customers served in a given period of time

2. 1/18 hour or 3.333 . . . minutes

3. $1/a$ hour

4. 1/20 hour or 3 minutes

5. $1/h$ hour

6. $x = a/h$; in this case, $x = 18/20 = 9/10$ or 0.9.

7. $L = 9$ customers; L is the average number of customers in the system at any given time.

8. $a =$ customers per hour; $L =$ customers; $W =$ hours

9. $W = 1/2$ hour; average time in the system is 1/2 hour, or 30 minutes.

10. $x = 16/20$ or 0.8; $L = 4$ customers; $W = 1/4$ hour or 15 minutes

11.

a	h	x	L	W
14	20	0.70	2.333	0.167
16	20	0.80	4.000	0.250
18	20	0.90	9.000	0.500
18	22	0.82	4.500	0.250
18	24	0.75	3.000	0.167

12. L and W increased.

13. L and W decreased.

14. $x = 1.1$, $L = -11$, $W = -0.5$; not all of these values make sense, because the number of customers and the wait time cannot be negative.

15. The average number of customers arriving is greater than the average number of customers that can be served in an hour. The number of people in line would increase indefinitely.

16. x is less than 1 everywhere in the table and is greater than 1 in question #14.

17. L is undefined. There is a vertical asymptote at $x = 1$.

18. All real numbers except for $x = 1$

19. You end up with $-1 = 0$, which is false. Therefore no solution exists at $L = -1$, which is a horizontal asymptote.

20. All real numbers except $L = -1$. The values of L are negative and decreasing. The number of customers in the system cannot be negative.

21. There are no customers arriving or waiting in the system. If $x < 0$, then $a < 0$ or $h < 0$. But a and h represent numbers of customers and must be nonnegative.

22. Students should trace the portion of the graph where $0 < x < 1$. $0 < x < 1$.

23. L increases rapidly (see table). The graph approaches positive infinity at $x = 1$.

x	L
0.8	4
0.9	9
0.95	19
0.99	99

24. The service rate, h

25. Get a faster server, or train the servers to increase their speed. (This would increase the value of h.) Add more ervers, and use a parallel-server model. (This would require a different model to calculate the average wait time for customers.) Some possible answers would be to provide entertainment (e.g., install a TV) and/or provide customers with serving numbers and a time to return.

Homework

1. **The queue at the W.C.** Students will be able to determine that the toilets are not fairly distributed. Because the arrival rate for men is 30 per hour and their service rate is 60 per hour, the traffic intensity for men is 30/60, or 50%. Therefore, for men, $L = .5/(1 - .5) = 1$, and $W = 1/30$ hour $= 2$ minutes. Thus, on average, a man spends 2 minutes in the system: 1 minute waiting and 1 minute using the facility. Because the arrival rate for women is also 30 per hour (assuming that the women arrivals are evenly distributed between the two women's toilets), the arrival rate for women is 15 per hour at each toilet. However, the service rate for women is only 20 per hour. Thus, the traffic intensity for women is 15/20, or 75%. Therefore, for women, $L = .75/(1 - .75) = 3$, and $W = 3/15$ hour $= 12$ minutes. So, on average, a woman spends 12 minutes in the system: 9 minutes waiting and 3 minutes using the facility.

➹ For more information and extension problems relating to the multi-server model, go to [**www.hsor.org**].

2. **The book checkout**
   a. On average, two people are in the system (both in line and being serviced).
   b. A person will spend a total of 3 minutes in line and checking out.
   c. A person will spend 2 minutes in line. Of the 3 minutes (on average) spent in the system, a person (on average) will spend 1 minute in service and 2 minutes waiting in line.

PROJECTS

All the projects give students field experience and exposure to the complexities of actual situations. To learn more about different service disciplines and arrival types for the first project, students will need to do some research.

➹ For more background information, students can go to [**www.hsor.org**].

TEACHER NOTES

Activity 7: Does This Line Ever Move?

Arm-and-a-Leg Ticket Sales

a (customers/hour)	h (customers/hour)	x	L (customers)	W (hours)
14	20			
16	20			
18	20			
18	22			
18	24			

Does This Line Ever Move? © 2005 Key Curriculum Press

Simulations

BACKGROUND

Simulation is the most flexible of all operations research modeling techniques. It can be used to model an existing system or one that is in the planning stages. Operations researchers often use a computer to generate random numbers that represent the elements of a situation or system. The analysis of the simulated data enables the manager to gain insights into the real-world system by manipulating the computer representation.

Flight simulators are used by civilian airlines and by the armed forces to train and test pilots. Business schools use simulations to teach students by providing them with simulated experiences of a wide range of business problems and opportunities. In each case, the value of simulation as a training tool lies in the ability to simulate a greater variety of potential situations than one would ever be likely to face in actual experience.

Simulations are also widely used to design a facility or develop operational policies. A computer simulation of a restaurant or manufacturing plant can be designed to evaluate different layouts and management policies, with the goal of reducing waiting time and increasing the rate of production. In general, these simulations are discrete-event **stochastic** simulations. That is, the computer simulation includes probability functions that are used to reflect randomness in the customer arrival pattern or the time needed to complete a task. The simulation model can be used to identify possible

Workers in a warehouse track inventory.

bottlenecks in a proposed system or to develop solutions for bottlenecks in an existing system.

Another area in which simulation is useful is inventory management. Companies that manufacture products or sell goods and services maintain inventory to meet the varying demands of the customer or the manufacturing process. The items in inventory may cost under a dollar each or be luxury cars worth $50,000 each or more. The item can be as small as a computer chip or as large as an airplane. It might be an engine or a box of cereal with a shelf life measured in years, or a bunch of grapes or a pint of donated blood with a shelf life of only days or weeks. In each case, someone must decide how many items of a specific type should be kept in inventory. If inventory is kept too low, it is possible that when a customer places a request there will be nothing in stock to sell. This unmet demand may represent a lost opportunity or simply added transportation expense if the item can be back-ordered. If inventory is too high and the product is perishable or seasonal, excess inventory may need to be thrown away or sold at a loss.

The seats on an airplane or rooms in a hotel present a different aspect of this same problem.

These items can be viewed as inventory that is subject to random demand and is used up each day. Any seat not sold on a given flight or hotel room not rented on a given night is inventory that is wasted. In this context, the inventory is essentially constant each day, and the challenge is to develop a pricing strategy that maximizes the total revenue by increasing demand. These strategies are sometimes called **yield management.**

CASE STUDY: SPORT OBERMEYER SKI WEAR

The fashion industry faces major challenges in predicting sales of specific product lines. The complexity of the problem is increased by variables such as color and size. Poor estimates can mean either an oversupply of inventory resulting in end-of-season price markdowns, or being out of stock at the end of the season resulting in customer frustration and lost sales.

Sport Obermeyer is a leading manufacturer of fashion skiwear. Their peak retailing season lasts two months, and demand is difficult to predict. As its business increased in the late 1980s, Obermeyer encountered serious manufacturing constraints during the critical summer months. At the same time, they needed to increase their variety in order to stay competitive. To improve forecasts, Sport Obermeyer took these actions.

- The twenty-five largest retail customers were invited to a sneak preview in February. Their early purchase orders were used to improve initial forecasts for the fall and winter.

- The members of the buying committee each made independent forecasts for different products. Products for which individual committee members differed the most were identified as having the most uncertain demand among customers. In this model, the data collected in-house was considered a simulation of what would take place in the market.

- Production capacity early in the season was optimized for products with more predictable demand. The mid-summer, critical production capacity was reserved for items with less predictable demand.

- Key design decisions reduced variability and cut the potential for component shortages holding up production. For example, increased use of black zippers as a color contrast reduced the number of zippers by a factor of 5.

- Colors and materials were standardized so that in a typical season, there were only two or three shades for each design cycle instead of five or six.

All of the bulleted actions were combined with sophisticated forecasting models, probabilistic inventory models, and optimal production plans. As a result, the need for markdowns was dramatically reduced in an industry in which profits average just 3% of sales. Costs declined by 2%, increasing profits to 5%.

References

Fisher, M. L., J. H. Hammond, W. Obermeyer, and A. Raman. 1994. Making supply meet demand in an uncertain world. *Harvard Business Review* May-June:83–92.

Fisher, M. L. and A. Raman. 1996. Reducing the cost of demand uncertainty through accurate response to early sales. *Operations Research* 44:87–99.

ACTIVITY **8**

Hot Dog Sales at Frankfurter High

Simulation can be used to decide how much of a product to have available for sale when demand is uncertain. This question is particularly important when dealing with perishable inventory, such as food, seasonal clothing, or newspapers, which lose value over time. In these cases, a decision must be made about how much to produce or buy in order to maximize profit.

The sports booster club at Felix J. Frankfurter High School, the Frankfurter "Hot Dogs," must decide how many hot dogs to purchase for sale at each basketball game. The team plays 10 home games each season, and every game is sold out. Hot dogs sell for $2.00 during games. The booster club buys hot dogs and buns by the dozen—one dozen hot dogs cost $6.56 and one dozen buns cost $1.44. At the end of every game, all unsold hot dogs and buns are donated to the local homeless shelter.

The Scenario

1. If the club sells one dozen hot dogs, how much money will it take in? How much profit will it make?

The Frankfurter "Hot Dogs" were optimistic about sales at the opening home game and ordered 16 dozen hot dogs and buns. However, they were able to sell only 9 dozen.

2. How much net profit did they make at the first game?

3. How much net profit would they have made if they had ordered exactly nine dozen hot dogs and buns?

4. How much lost profit resulted from over-ordering? What conclusion might you draw about future purchases?

Because they had leftovers, the Frankfurter "Hot Dogs" decided to order only nine dozen hot dogs and buns for the next game. However, this game drew a large, hungry crowd, and they ran out of hot dogs! They estimated they could have sold another five dozen hot dogs.

5. How much profit did the booster club make at the second game?

6. How much profit would it have made if it had ordered and sold another five dozen hot dogs?

7. How much potential profit was lost due to under-ordering? What conclusion might you now draw about future purchases?

8. How many dozen would you choose to order for the next game? How much confidence do you have in your choice? Why?

Developing a Model

The members of the booster club are frustrated. They realize that the demand for hot dogs is random, so they can't predict exactly how many hot dogs they will be able to sell at any specific home game. They decide to analyze their records for the past five years. They learn that

- They always ordered 16 dozen hot dogs and rolls.
- The average demand was for around 10 or 11 dozen hot dogs.
- The most they ever sold was 16 dozen.
- The least they ever sold was 5 dozen.
- The demand for hot dogs frequently was between 9 and 12 dozen.
- The demand was rarely less than 9 dozen or more than 12 dozen.

They decide to create a simulation to model the problem, so that they can analyze potential costs and profits over a period of time without spending any money. They use three dice with six faces numbered 1 through 6. The sum of the numbers on the three dice, a whole number between 3 and 18, inclusive, will represent the number of dozens of hot dogs sold. Will the average result of rolling three dice accurately model the average demand for hot dogs?

9. Because each of the six numbers on each die is equally likely to occur, the long-term average for a single die is just the average of the six numbers. What is the long-term average of rolling a single die with six faces numbered from 1 through 6?

10. What is the long-term average of rolling three dice?

11. Does rolling three dice accurately model the demand for dozens of hot dogs at Frankfurter High's home basketball games? Explain.

Performing the Simulation

To simulate the demand for hot dogs at Frankfurter High's home basketball games, work with a partner and a set of 3 dice. Use each roll of the 3 dice to represent the demand, in dozens of hot dogs, for one home game. Simulate the demand for one season by rolling the dice 10 times.

12. With your partner, decide on an amount of hot dogs and buns, in dozens, to order every week, which you think might maximize profit.

13. Calculate the total weekly fixed cost for the number of dozens of hot dogs and buns you decided to order. Enter that value for weeks 1 through 10 in the column labeled "Weekly Fixed Cost" in a copy of this table.

Week	Your New Policy: Always Order ____ Dozen					Old Policy: Always Order 16 Dozen		
	Random Demand (dozens)	Hot Dogs Sold (dozens)	Total Revenue	Weekly Fixed Cost	Profit = Revenue − Cost	Hot Dogs Sold (dozens)	Total Revenue	Profit = Revenue − $128
1								
2								
⋮								
9								
10								
Totals								
Average								
Min								
Max								

14. Now perform the simulation for one season and complete your table. For each game, record the sum of the three dice in the column labeled "Random Demand." Compare the demand to your weekly order and record the number of dozens of hot dogs sold. (You can't sell more hot dogs than you have available!) Then compute the total revenue and profit for each game. Complete the last three columns for the old policy of always ordering 16 dozen at a cost of $128 per game. Finally, compute the total and average for each column.

15. What was the least profit you made for any game? What was the most profit you made for any game? What caused this difference?

16. How does the total profit for the season under the old policy compare to the total profit under your new policy?

A Second Simulation

Do you think that every one-season simulation would produce approximately the same results? You'll explore this issue by conducting another one-season simulation with the same ordering policy used in the previous section.

17. Perform the same procedures to carry out the simulation a second time. Record your results in another copy of the table.

18. What was the least profit you made for any game? What was the most profit you made for any game?

19. Compare the results of the two simulations. Are they the same? Why or why not?

20. Do you think the old policy is better, or the new one?

Finding the Relative Frequency

Next, we will use **observed relative frequency** to estimate the probability of each demand. Recall that **relative frequency** is the proportion of times that a specific outcome occurs.

21. The possible demand outcomes simulated by tossing three dice are the integers from 3 through 18. For each of your two simulations, record the frequency of each demand possibility in a copy of the table below. Calculate the relative frequencies by dividing each frequency by 10. Finally, pool the results of all the simulations performed by your class, calculate the relative frequencies, and record the values in your table.

Demand Values	First Individual Frequency (i)	First Relative Frequency ($i/10$)	Second Individual Frequency (j)	Second Relative Frequency ($j/10$)	Class-Pooled Frequency (p)	Class-Pooled Relative Frequency
3						
4						
5						
⋮						
17						
18						
Min						
Max						

22. Using your graphing calculator, create histograms of the two individual relative frequencies and of the pooled relative frequency. Describe the shape of each of the three histograms.

23. Are any of the histograms symmetrical? If so, which histogram is most symmetrical? Why does this make sense?

24. Which value occurred most frequently in the individual results? In the pooled results?

25. As a class, choose any one of the possible demand numbers, and record that choice. For each of your individual simulations, what is the relative frequency of the demand number you chose?

26. For the class-pooled simulations, what is the relative frequency of the demand number you chose in question #25?

27. Identify the pair of students in your class who recorded the *largest* relative frequency of the demand number chosen by the class. What is the relative frequency of that number? Next, identify the pair who recorded the *smallest* relative frequency of that demand number. What is that smallest relative frequency?

28. What is the order relationship of the three relative frequencies from questions #26 and #27? Write an inequality to illustrate this order relationship.

29. What was the greatest total profit for a season simulated by a pair of students in your class? Which ordering policy resulted in this profit? What was the average demand for this simulated season?

Finding the Best Policy

Finally, we will determine the best long-term ordering policy by using the pooled relative frequency data to simulate long-term random demand. In order to evaluate every possible ordering policy, your teacher will assign to you an ordering policy somewhere between 3 and 18 dozen, inclusive.

30. Enter your assigned policy in a copy of the table below. Complete the first four columns of your table. The pooled relative frequency for each demand in the simulations approximates the proportion of games having that demand in the long term. Multiply the profit by the relative frequency and sum that column to determine a weighted average profit per game under your assigned policy.

	Your Assigned Policy: Always Order _____ Dozen				
Demand	Hot Dogs Sold	Total Revenue	Profit	Pooled Relative Frequency	Profit × Pooled Relative Frequency
3					
4					
5					
⋮					
17					
18					
				Total (Weighted Average)	

31. Compare the weighted average for each possible policy calculated by your classmates. What is the best ordering policy? What is its associated weighted average profit?

32. Compare your answers to those for questions #27 and #29 and explain any differences.

Everyday Applications of Operations Research

DINAMO Yield Management at American Airlines Has a $1.4 Billion Impact

In 1968, American Airlines launched its first automated overbooking process. Without overbooking, American estimates that 15% of the seats on sold-out flights would be unused, due to cancellations. Forecasting demand and cancellation rates was the first element in developing a yield management system. The present system, called DINAMO, is a sophisticated yield management system credited with saving American $1.4 billion over a three-year period in the 1990s.

➤ For more case studies and background, go to (**www.hsor.org**).

EXTENSION 1: WHAT ARE THE POSSIBILITIES?

To gain a better understanding of the hot dog sales simulation, you'll analyze the entire **probability space** for tossing three dice. The probability space is the set of all possible outcomes. For example, the probability space for tossing one die is 1, 2, 3, 4, 5, 6.

1. When *one* die is tossed, how many outcomes are possible? When two are tossed? When three are tossed?

2. Write two possible outcomes of tossing three dice, in the form (x, y, z). What systematic strategy could you use to list every outcome and be sure you didn't miss any?

3. How many ordered triples in the probability space have the form $(1, 1, z)$? List them all, and write the *sum* of the three dice for each outcome.

4. Repeat question #3 for all the ordered triples of the form $(1, 2, z)$. Then do the same for ordered triples of the form $(1, 3, z)$ through $(6, 6, z)$.

5. For how many of the ordered pairs from questions #3 and #4 is the sum of the dice 3?

6. For how many of the ordered pairs from questions #3 and #4 is the sum 18?

7. What is the probability of obtaining a sum of 3? Of 18? Write your answers in decimal form.

8. Find the probabilities of obtaining sums of 4, 5, and so on, through 17.
   Write your answers in decimal form.

9. How do the probabilities you found in questions #7 and #8 compare to the
   pooled relative frequencies from the hot dog sales activity?

➤ For graphing calculator simulations and extensions, go to [**www.hsor.org**].

HOMEWORK

1. **Salvage value.** The Frankfurter "Hot Dogs" decide to sell the unsold hot
   dogs and buns at a deep discount to fans as they leave. Assume that they will
   be able to sell all of the leftover dozens at a price of $5 per dozen. Use the
   tables below to reevaluate the policy of ordering 11, 12, or 13 dozen with this
   "salvage" value of $5 per dozen. For the column "Pooled Relative Frequency,"
   use the data generated earlier in class. For example, if you ordered 11 dozen
   and the demand was only 9 dozen, there would be 2 dozen to sell at a
   discount. The total net revenue would increase by 2($5), or $10.

   a. Copy and complete this table for ordering policies of 12 dozen, 13 dozen,
      14 dozen, and 15 dozen.

	Always Order _____ Dozen							
		No Salvage Value		$5/dozen Salvage Value			No Salvage Value	$5/dozen Salvage Value
Demand	Hot Dogs Sold	Total Revenue	Profit A	Total Revenue	Profit B	Pooled Relative Frequency	Profit A × Pooled Relative Frequency	Profit B × Pooled Relative Frequency
3								
4								
⋮								
17								
18								
						Totals (Weighted Average)		

   b. What is the best policy with no salvage value?
   c. What is the best policy with a $5 salvage value?

PROJECT: MANAGING INVENTORY

Interview a business manager and/or collect data from an actual business to determine

- daily (or weekly) demand for a specific product
- store policy on how many to order and when
- store policy regarding unsold product (for example, discounted, returned, thrown away)

You can look at products with varying periods of analysis: daily demand (for example, fresh bread or newspapers), weekly demand (for example, milk or magazines), seasonal demand (for example, clothing or garden supplies), and rental items (for example, specific videos or cars).

Hot Dog Sales at Frankfurter High

This lesson demonstrates for students a process that brings the power of probabilistic reasoning to a difficult decision involving uncertainty. Students are introduced to the concept of random demand and the use of simulation to identify the best alternative.

OBJECTIVES

- experience randomness and the law of large numbers
- relate long-term averages of a random variable to expected value
- experience business decisions made in the face of uncertainty
- experience the effect of pooling results

TECHNOLOGY NOTE

Students can use a spreadsheet program to peform the various calculations.

PRE-ACTIVITY: THE LAW OF LARGE NUMBERS *(optional)*

As a warm-up, you can use this activity to set up the concept of a long-term average result when dice are tossed.

1. Imagine tossing an ordinary fair die 100 times. What do you predict the average of these 100 tosses would be?

2. Toss a die 100 times, record the results, and compute the average outcome.

Guiding the Pre-activity

After students have completed steps 1 and 2, discuss these ideas:

- The long-term average does not need to be a number that is actually obtainable.

- Compare the range of values obtained by pairs of students for the average of 100 tosses. Why are the averages not the same? (The standard deviation of the average of a sample of 100 tosses is 0.17. Ninety-nine percent of the observed averages should be within 3 standard deviations of the mean.

Therefore, virtually all of the observed averages should fall between 3 and 4. Most likely, they will fall between 3.16 and 3.84, which are two standard deviations around the mean.)

- What number makes sense as the theoretical long-term average? Why does this number make sense? (Because each of the numbers on a fair die is equally likely, the long-term average must be the average of the six numbers: $(1 + 2 + 3 + 4 + 5 + 6)/6 = 21/6 = 3.5$. If students have experience with expected value, you can connect the concepts of expected value and long-term average.)

GUIDING THE ACTIVITY

Questions #1–13 set the scenario and are intended to point out two potential inventory problems—over-ordering (causing some loss of profit) and under-ordering (leading to loss of potential profit).

You may want to motivate the concept of a simulation by asking students why using the results of a simulation, rather than actual results, makes sense. You might mention flight simulators as an example. Flight simulators are used to train pilots to respond appropriately to in-flight problems, because the use of a flight simulator does not put an expensive aircraft or the pilot at risk. In the hot dog sales activity, it would simply take too long and cost too much to accumulate enough real data.

The simulation in the activity uses three dice because the possible outcomes are integers between 3 and 18, inclusive. The extremes in demand given in the activity are close to the extreme values obtainable when three dice are rolled. However, you may need to point out that although demand for 3, 4, 17, and 18 dozen did not occur in the past five years, such demands could occur in the future. You may wish to discuss the fact that the sums of two or more dice are not equally likely, and that those totals in the middle of the range are much more likely than the extremes.

For question #21, you will need to facilitate collection of the class-pooled data. Remember that each pair of students has completed 20 simulations. Students will need to think about how to calculate the class-pooled relative frequency. They will need to divide each pooled frequency by 10 times the number of students in the class.

Class-pooled relative frequency data should be close to the theoretical probabilities given in the table shown.

Demand	Probability	Demand	Probability
3	.005	11	.125
4	.014	12	.116
5	.028	13	.097
6	.046	14	.069
7	.069	15	.046
8	.097	16	.028
9	.116	17	.014
10	.125	18	.005

(Students will calculate these theoretical probabilities in Extension 1.)

SOLUTIONS

Activity

1. $24; $24 − ($6.56 + $1.44) = $16

2. $24(9) − $8(16) = $216 − 128 = $88

3. $16(9) = $144

4. $56; to maximize profit, it is important not to over-order.

5. $16(9) = $144   6. $16(14) = $224

7. $80; to maximize profit, it is important not to under-order.

8. Answers will vary, but it is difficult to have much confidence in any answer, in the face of uncertain demand.

9. (1 + 2 + 3 + 4 + 5 + 6)/6 = 3.5

10. 3(3.5) = 10.5

11. Based on the research on prior years' sales, the extreme values are close to the highest and lowest possible values of the sum total of three dice. The average demand is close to the expected value of the sum of three dice. The averages are close, and the numbers in the middle of the range occur more frequently, as was true for the demand for hot dogs.

12. Answers will vary, hopefully between 3 and 18!

13. Answers will vary depending on the response to question #12.

14. Answers will vary.

15. Answers will vary. The difference in profit is caused by the randomness in the simulated demand. The most profit will correspond to rolls of the dice that meet or exceed the ordering policy. You may want to point out that the number of hot dogs "actually sold" is truncated by the order policy. The least profit occurs when the demand is low, which corresponds to a low roll of the dice. Some students may get a negative profit (a loss) if they ordered a large number but demand happened to be small.

16. Most likely the profit will be higher under the new policy. However, it is possible that if a student chose an extremely small number to order, the profit would actually go down as compared to always ordering 16 dozen. You may want students also to make a comparison with a policy that almost always represents an under-order and guarantees that nothing is left over. For example, order 9 dozen every week.

17. and 18. Answers will vary.

19. Almost certainly no; this is due to the rolling of the dice that simulates randomness in the demand.

20. and 21. Answers will vary.

22. The first two histograms will probably not show much shape, because there were only 10 simulations in each, but the pooled data should approximate a normal distribution.

23. There may not be any apparent symmetry in the first two histograms, due to the small number of simulations, but the histogram of the pooled data should at least hint at symmetry about 10 and 11. (Recall that the "average" sum of three dice is 10.5.) The pooled data most likely will show more symmetry than any of the individual histograms.

24.–27. Answers will vary.

28. No matter which demand number was chosen, the smallest individual relative frequency will be less than the pooled relative frequency, and the largest individual relative frequency will be greater than the pooled relative frequency. Smallest individual relative frequency < pooled relative frequency < largest individual relative frequency. This illustrates the "law of large numbers" at work. The larger the sample size (as in the pooled results), the more nearly the empirical (or experimental) results will mirror the theoretical results.

29. Answers will vary, but beware of an unusually high or low profit number due to miscalculation. Almost all total profits should be in the range of $1100 to $1600. The highest profit any student

obtains is related to both ordering policy and relatively high average demand.

30. Answers will vary.

31. Answers will vary. The best student individual performance depends on both the policy selected and the randomness of the die. Based on the expected value, the best long-term policy is 12 (net profit of $142.00 per game), with a policy of 11 only slightly worse (net profit of $141.00 per game) and 13 a close third (net profit of $140.22 per game). In all likelihood, the student with the best results will have used one of these three ordering policies.

32. Answers will vary, but students may cite greater confidence when more data were used.

Extension 1: What Are the Possibilities?

1. When *one* die is tossed, 6; when *two* are tossed, $6^2 = 36$; when *three* are tossed, $6^3 = 216$

2. Answers will vary, e.g., (1, 2, 3); hopefully some students will see that one way to list all outcomes systematically is to hold two of the numbers constant and let the third number vary across the six possibilities.

3. The six possibilities and their totals are (1, 1, 1), 3; (1, 1, 2), 4; (1, 1, 3), 5; (1, 1, 4), 6; (1, 1, 5), 7; (1, 1, 6), 8.

4. (1, 2, 1), 4; (1, 2, 2), 5; (1, 2, 3), 6; (1, 2, 4), 7; (1, 2, 5), 8; (1, 2, 6), 9; (1, 3, 1), 5; . . . ; (1, 3, 6), 10; (1, 4, 1), 6; . . . ; (1, 4, 6), 11; (1, 5, 1), 7; . . . ; (1, 5, 6), 12; (1, 6, 1), 8; . . . ; (1, 6, 6), 13; (2, 1, 1), 4; . . . ; (2, 1, 6), 9; (2, 2, 1), 5; . . . ; (2, 2, 6), 10; (2, 3, 1), 6; . . . ; (2, 3, 6), 11; (2, 4, 1), 7; . . . ; (2, 4, 6), 12; (2, 5, 1), 8; . . . ; (2, 5, 6), 13; (2, 6, 1), 9; . . . ; (2, 6, 6), 14; . . . ; (3, 1, 1), 5; . . . ; (3, 1, 6), 10; (3, 2, 1), 6; . . . ; (3, 2, 6), 11; (3, 3, 1), 7; . . . ; (3, 3, 6), 12; (3, 4, 1), 8; . . . ; (3, 4, 6), 13; (3, 5, 1), 9; . . . ; (3, 5, 6), 14; (3, 6, 1), 10; . . . ; (3, 6, 6), 15; (4, 1, 1), 6; . . . ; (4, 1, 6), 11; (4, 2, 1), 7; . . . ; (4, 2, 6), 12; (4, 3, 1), 8; . . . ; (4, 3, 6), 13; (4, 4, 1), 9; . . . ; (4, 4, 6), 14; (4, 5, 1), 10; . . . ; (4, 5, 6), 15; (4, 6, 1), 11; . . . ; (4, 6, 6), 16; (5, 1, 1), 7; . . . ; (5, 1, 6), 12; (5, 2, 1), 8; . . . ; (5, 2, 6), 13;

(5, 3, 1), 9; . . . ; (5, 3, 6), 14; (5, 4, 1), 10; . . . ; (5, 4, 6), 15; (5, 5, 1), 11; . . . ; (5, 5, 6), 16; (5, 6, 1), 12; . . . ; (5, 6, 6), 17; (6, 1, 1), 8; . . . ; (6, 1, 6), 13; (6, 2, 1), 9; . . . ; (6, 2, 6), 14; (6, 3, 1), 10; . . . ; (6, 3, 6), 15; (6, 4, 1), 11; . . . ; (6, 4, 6), 16; (6, 5, 1), 12; . . . ; (6, 5, 6), 17; (6, 6, 1), 13; . . . ; (6, 6, 6), 18.

5. 1   6. 1

7. 1/216 in each case; ≈.005

8. $P(4) \approx P(17) \approx 3/216 \approx .014$; $P(5) \approx P(16) \approx 6/216 \approx .028$; $P(6) \approx P(15) \approx 10/216 \approx .046$; $P(7) \approx P(14) \approx 15/216 \approx .069$; $P(8) \approx P(13) \approx 21/216 \approx .097$; $P(9) \approx P(12) \approx 25/216 \approx .116$; $P(10) \approx P(11) \approx 27/216 = .125$

9. Answers will vary.

Homework

1. **Salvage value**
   a. Answers will vary.
   b. The answers will vary depending upon the results of your pooled relative frequency. We used the actual probabilities and found the optimal policy was to order 12 dozen with an expected value of $142.00 per week.
   c. The answers will vary depending upon the results of your pooled relative frequency. We used the actual probabilities and found the optimal policy was to order 14 dozen with an expected value of $154.42 per week. The table below represents the expected values based on probabilities. You should expect similar values, but due to randomness, the optimal strategy may differ.

	Order 11	Order 12	Order 13	Order 14
Salvage $0	$141.00	$142.00	$140.22	$136.11
Salvage $5	$148.79	$152.42	$154.42	$154.42

TEACHER NOTES

Activity 8: Hot Dog Sales at Frankfurter High

Hot Dog Sales Simulation

Week	Random Demand (dozens)	Your New Policy: Always Order ___ Dozen				Old Policy: Always Order 16 Dozen		
		Hot Dogs Sold (dozens)	Total Revenue	Weekly Fixed Cost	Profit = Revenue − Cost	Hot Dogs Sold (dozens)	Total Revenue	Profit = Revenue − $128
1								
2								
3								
4								
5								
6								
7								
8								
9								
10								
Totals								
Average								
Min								
Max								

Activity 8: Hot Dog Sales at Frankfurter High

The Relative Frequency of Different Demands

Demand Values	First Individual Frequency (i)	First Relative Frequency ($i/10$)	Second Individual Frequency (j)	Second Relative Frequency ($j/10$)	Class-Pooled Frequency (p)	Class-Pooled Relative Frequency
3						
4						
5						
6						
7						
8						
9						
10						
11						
12						
13						
14						
15						
16						
17						
18						
Min						
Max						

Activity 8: Hot Dog Sales at Frankfurter High

Finding the Best Policy

Your Assigned Policy: Always Order ___ Dozen					
Demand	Hot Dogs Sold	Total Revenue	Profit	Pooled Relative Frequency	Profit × Pooled Relative Frequency
3					
4					
5					
6					
7					
8					
9					
10					
11					
12					
13					
14					
15					
16					
17					
18					
		Total (Weighted Average)			

ACTIVITY 9

Torn Shirts, Inc.

PART ONE: USING DATA TO ANALYZE A BUSINESS PROBLEM

Torn Shirts, Inc. (TSI), specializes in silk-screened shirts "ripped" to order. Three high school students, Marlene, Elina, and Hong, run the business. They have come up with a variety of sayings, such as "I've been ripped off!" and their shirts have become very popular. The potential buying population is the 10,000 high school students in their local district.

THE SCENARIO

TSI distributes a bimonthly flyer announcing their latest items. The weekly fixed cost for marketing, telephone lines, and the computer system averages $100. The net revenue per shirt, excluding the weekly fixed cost, is $6.

Computer programs can simulate a hypothetical scenario, such as the location of a new ballpark, providing users a chance to explore some of the unknowns, such as the effects on traffic and neighboring businesses.

1. How many shirts do they need to sell in a week to cover their fixed costs?

2. Each of the three partners would like to earn $100 in profit per week to make their participation in the business worthwhile. How many shirts would they need to sell to achieve their profit goal and cover their costs?

The Call Log

The owners take turns taking orders by telephone Monday through Friday between the hours of 6 P.M. and 9 P.M. Some of their friends say that they frequently get a busy signal when they call to place an order, and so they don't call back. Marlene and Elina have decided to track all calls for a typical week. If they are missing too many calls, they need to consider getting call waiting or installing another telephone line.

For each call answered, they record the time of the call and the duration of the call in minutes. The data for the week are shown in these tables.

ANSWERED CALLS FOR A TYPICAL WEEK

Monday		Tuesday		Wednesday		Thursday		Friday	
Time	Duration	Time	Duration	Time	Duration	Time	Duration	Time	Duration
6:09.9	1.3	6:00.3	4.1	6:11.8	7.6	6:05.1	1.7	6:01.4	7.3
6:36.8	23.0	6:17.4	1.5	6:22.1	9.1	6:14.8	1.2	6:24.4	9.8
7:07.0	2.2	6:36.6	4.8	6:31.3	3.5	6:22.0	18.1	6:39.7	3.6
7:23.4	12.9	6:54.9	2.9	6:52.4	1.7	6:46.2	11.2	6:45.7	4.5
7:45.3	8.3	7:03.6	12.6	6:57.7	16.2	7:10.2	2.3	6:53.9	15.7
8:22.0	8.1	7:21.2	2.8	7:14.2	1.5	7:33.4	4.3	7:35.2	1.4
8:30.9	1.8	7:26.6	4.0	7:18.5	4.7	7:38.3	4.0	7:48.3	2.3
8:38.5	1.8	7:30.9	12.6	7:24.5	4.2	7:43.5	20.0	7:50.9	1.7
		8:12.2	7.6	7:29.7	11.0	8:06.9	1.2	8:09.1	7.8
		8:28.7	19.6	8:19.1	5.4	8:29.9	1.2	8:23.7	7.6
		8:56.2	3.1	8:30.3	4.3			8:51.2	15.1
				8:52.9	2.9				

3. How many shirts were ordered during the week, if the average order per call is one shirt?

4. Find the total profit for the week.

5. Did the girls meet their profit goal?

6. Find the average number of calls per hour for the week, recalling that phones are staffed three hours per night.

7. Find the total time spent answering calls.

8. Find the mean duration of an answered call.

The Probability of Getting a Busy Signal

Elina realized she could determine the probability that a caller gets a busy signal by using the data. Assuming that the calls arrive at random, this probability is equal to the fraction of time that the line is busy between 6 P.M. and 9 P.M. Any time spent talking to a customer after 9 P.M. must be subtracted from the total number of minutes on the phone.

9. Determine the total number of minutes spent talking to a customer after 9 P.M.

10. What is the total number of minutes per week that the line is staffed?

11. What is the probability, as a percentage, of a caller getting a busy signal?

12. Explain the significance of the percentage of time the phone is busy in terms of its impact on customer service.

13. If we assume that customers do not call back when they get a busy signal, what percentage of their potential customers are Marlene, Elina, and Hong missing?

14. Use the answer to question #13 to find the percentage of incoming calls that were answered.

15. Estimate the total number of calls during the week, including those that were missed.

16. Use the information in questions #3 and #15 to estimate the number of calls that were missed and the potential revenue from them.

Trends in the Call Data

Marlene, Elina, and Hong want to get a sense of the timing and duration of the calls they answer. They graph the data and look for trends.

17. Make a histogram that shows the number of answered calls in each half hour of the evening. Combine all the data for the week. Mark the horizontal axis in 30-minute intervals (starting at 6:00), and mark the vertical axis with the frequency of calls during each half hour, as shown here.

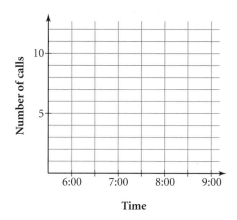

18. Look for patterns in your histogram from question #17, and describe its general shape.

19. Based on your histogram, which of these options is most likely?

   a. A given call is received before 7:30.0.
   b. A given call is received after 7:30.0.
   c. Choice (a) is just as likely as choice (b).

20. Make a histogram of the time spent on each answered call for the entire week. Mark the horizontal axis in 5-minute intervals (0 to 5, 5.1 to 10, and so on), and mark the vertical axis with the frequency of calls whose duration falls within that time interval, as shown here.

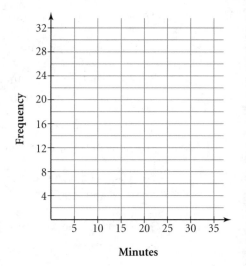

21. Look for patterns in your histogram and describe its general shape.

22. Draw a star on the horizontal axis of your histogram at the average duration of a call. (You calculated this in question #8.) Based on your histogram, which of these options is most likely?

    a. A given call is longer than the average call.
    b. A given call is shorter than the average call.
    c. Choice (a) is just as likely as choice (b).

23. Arrange the duration of the 52 calls in order from the shortest to the longest. Find the median duration of a call.

24. Now consider the median to be the "average." Based on the data, which of the options listed in question #22 (a, b, or c) is more likely?

25. Your histogram from question #20 should show a skewed-right distribution. How are the mean and median related in this type of distribution?

PART TWO: ANALYZING SIMULATED DATA TO MANAGE TELEPHONE ORDERS

The owners of Torn Shirts, Inc., now realize that they are losing a large number of incoming calls. They would like to increase the number of calls that can be answered, but they are unsure whether the solution is to add a phone line or a call-waiting option. Elina's father, an operations researcher, suggests evaluating the different options using data generated by a computer simulation of incoming calls. The computer simulation uses the trends in the call data to generate a random pattern of calls that is consistent with the averages. The data appear in the table Simulated Call Data for Monday. In this part of the activity, you'll help TSI analyze the possible increase in the number of answered calls if (a) they buy call waiting, which allows one call to be placed on hold, or (b) they install a second call line and staff both phones.

One Line with No Call Waiting

Refer to the table Simulated Call Data for Monday. Complete the column labeled "One Line (no call waiting)" first. Record the time each call ends, if answered. Identify missed calls with an "M."

SIMULATED CALL DATA FOR MONDAY

Call #	Time of Call	Duration, Actual or Potential (in minutes)	One Line (no call waiting)		One Line (with call waiting)			Two Lines (no call waiting)		
			End Time	M	End Time	Wait Time	M	End Time 1*	End Time 2	M
1	6:09.9	1.3	6:11.2		6:11.2			6:11.2		
2	6:36.8	23.0	6:59.8		6:59.8			6:59.8		
3	6:38.2	1.3		M	7:01.1	21.6			6:39.5	
4	6:42.2	6.6		M			M		6:48.8	
5	6:45.3	2.2		M			M			M
6	6:59.0	12.1		M			M		7:11.1	
7	6:59.4	15.0		M			M			M
8	7:07.0	2.2	7:09.2		7:09.2			7:09.2		
9	7:23.4	12.9								
10	7:27.3	14.4								
11	7:45.3	8.3								
12	7:49.0	1.9								
13	7:50.1	2.7								
14	8:22.0	8.1								
15	8:30.9	1.8								
16	8:38.5	1.8								

End time = completion time for calls answered immediately
M = missed call
Wait time = waiting time for calls experiencing call waiting, eventually answered
End time 1 = completion time for calls answered immediately on line 1*
End time 2 = completion time for calls answered immediately on line 2

*Default to line 1 when both lines are open.

1. What must occur in order for a call to be answered?

2. Explain why calls are missed. Should missed calls have a duration time? Why or why not?

3. What is the total number of missed calls?

One Line with Call Waiting

Continue to work with your table of Simulated Call Data for Monday. Answer questions #6–16 as you complete the column labeled "One Line (with call waiting)."

4. How long would call #3 wait to be answered? Why would it not be answered immediately? Record this waiting time in the column labeled "Wait Time."

5. Explain the end time, 7:01.1, for call #3.

6. Explain why none of calls #4 through #7 would be placed on call waiting.

7. For calls #9 through #16, complete the columns labeled "End Time," "Wait Time," and "M." (Remember to check the end time of two previous calls in order to determine whether a call would be missed.) Would call #9, at 7:23.4, be answered? If so, when would call #9 end?

8. How long would call #10 have to wait?

9. Would call #11 be answered immediately, placed on call waiting, or missed? (Hint: Refer to the end time of calls #9 and #10!) If call #11 were answered, at what time would the call end?

10. In this simulation, would TSI answer any more calls if they had call waiting? Would they still have any missed calls? Explain.

11. What is the total number of Monday callers placed on call waiting? What is the average time these callers waited? Do you think callers would be willing to wait this long?

Two Lines Without Call Waiting

Complete your table of Simulated Call Data for Monday by answering questions #12–19 as you complete the column labeled "Two Lines (no call waiting)." Assume that line 2 will be used only when line 1 is busy.

12. Why would the end time for call #3 not be the same for one line (with call waiting) as for two lines (no call waiting)?

13. Explain why a call could still be missed with two phone lines.

14. For calls #9 through #16, complete the columns labeled "End Time 1," "End Time 2," and "M." How many calls would be answered on line 2?

Finding the Best Option

Now let's analyze which of the three options is best.

15. Record the number of answered calls for each of the three options: one line (no call waiting), one line (with call waiting), and two lines (no call waiting).

16. Calculate the total net revenue, excluding fixed costs, for each of the three options.

17. The cost of call waiting is an additional $10 per week. The cost of a second business line is $15 per week. Which of the three options would you recommend to TSI? Why?

18. Will the two-line option enable the three business partners to meet their goal of earning $100 per week each? Explain. Why might the $100 goal for each partner change under the option of two lines?

19. Should TSI make this decision based on this Monday's data? Explain.

Everyday Applications of Operations Research

AT&T Increases Market Share by $1 Billion— Call Processing Simulator Boosts Profits

AT&T reports that its Call Processing Simulator, a computer program used to analyze possible changes in the design and operation of call centers for customers, has helped increase its share of the $8 billion call-center service industry by $1 billion. In 1992 alone, AT&T conducted over 2,000 simulation studies for its customers. AT&T customers implementing the recommendations resulting from these simulation studies increased their annual profits by $750 million. For example, when Northwest Airlines implemented the results recommended by the simulation, it increased its number of calls answered by 20%, with 20% fewer agents and 27% less in overtime costs. This resulted in a 5% increase in revenue for the airline.

➤ For more case studies and background, go to ⟨ **www.hsor.org** ⟩.

PROJECT: SIMULATION IN THE WORKPLACE

Write a letter to a business or government agency asking how simulation is used to improve operations.

Torn Shirts, Inc.

This is a simplified version of a call-center simulation, similar to the simulations used by businesses to improve their operations and revenue. Students will use simulated data to determine the best of three phone-service options for a small company.

OBJECTIVES

- calculate measures of central tendency
- make and interpret histograms
- analyze and manipulate tables of data
- interpret data in a business context
- make decisions based on analysis of simulated data

PRE-ACTIVITY: DATA SIMULATION

You can use this pre-activity to give students a feel for data simulation.

Suppose a small-business owner receives calls from 6 P.M. to 9 P.M. Use the random number generator on your calculator to generate and record call times and durations. Record them in a table like the one shown here.

Call Time	Duration	End Time
⋮		

- Generate a random integer between 0 and 30 (minutes) and add it to 6 P.M. Record the time in the "Call Time" column as the first call time.

- Generate a second random integer between 0 and 30 (minutes) and add it to the previous call time and record it in the table as the second call time. Repeat this process until the call time is after 9 P.M. Stop and do not add that call time to the table.

- Begin with the first call and use the random number generator to generate a random integer between 0 and 20 (minutes). Record this number in the "Duration" column. Calculate the end time of the first call and record it in the last column.

- Compare the end time of the first call and the call time of the second call to determine whether or not the second call would have been answered. If yes, generate a call duration for this second call and record the data. If no, record "None" in the second column. Repeat the process until every call either has a duration time or the word "None" in the second column. Then have students compare their results and answer these questions.

1. How many calls were made? (Answers will vary; 12, on average.)

2. How many calls were answered? What proportion of calls were unanswered? (Answers will vary; on average, over 60% answered.)

3. What is the average number of calls per hour? What is the average duration? (4; 10 minutes)

Guiding the Pre-activity

This activity will familiarize students with the process of simulating a call center. Close the activity by asking, "If this were real rather than simulated data, what information in the table would be unobservable?" Students should realize that it is difficult for a business to determine the number of missed calls. Two possible ways of estimating the number of missed calls are to

- randomly call the business from the outside

- determine the percentage of time the line is busy

In the first part of the student activity, students develop the latter approach. You might conclude with a discussion of how businesses alleviate these problems.

GUIDING THE ACTIVITY

The activity requires a large amount of repetitive arithmetic. We recommend that students use calculators or spreadsheets and work as teams so that they can share the work and check each other to avoid careless errors. Simulated call data for Tuesday through Friday are included. You may choose to print these out and assign different groups to analyze different days.

Part One: Using Data to Analyze a Business Problem

You might begin with a class discussion of personal experiences in telephoning a business and being placed on hold or getting a busy signal. How often does it happen? How long do you wait on hold? Do you ever hang up? If you hang up, do you call back, or do you look elsewhere for the same product or service?

You could continue with a discussion of options that a business has to improve service or avoid losing calls and the advantages and disadvantages of each option. You might discuss what types of phone service options are available, including call waiting, multiple lines, and voice mail. Finally, you might ask how the Internet affects such businesses.

You may wish to use the pre-activity, especially if your students are unfamiliar with the mechanics of simulation.

The times in the tables give decimal fractions of a minute. For instance, 6:30.2 represents 6:30:12.

Every mathematical model of a real-world situation is based on a particular set of assumptions. The simulation model in this lesson uses five assumptions: the pattern of call arrivals is random; the time required to service a call is also random; each caller will order exactly one shirt; callers who receive a busy signal will not call back; and callers placed on call waiting will remain on the line until their call is answered. You may wish to inform students of these assumptions and discuss their validity.

As students approach question #17, you may wish to review how to make a histogram as well as terminology for describing its shape. In a uniform distribution, each interval has the same frequency. A normal, or "bell-shaped," distribution is symmetric, with the largest frequency in the middle interval and decreasing frequencies for intervals that are successively farther from the middle. If a distribution is skewed to the right or left, then the largest frequencies will be left or right, respectively, of the middle interval.

Part Two: Analyzing Simulated Data to Manage Telephone Orders

You may want to conclude the class with a discussion of question #19, which addresses two important aspects of the simulation. First, there might be some variation in the number of calls across the days of the week. Second, the simulation should be performed a number of times, and long-term averages should be used as the basis for any decision.

EXTENSION

You can ask students to analyze the simulated data for other days of the week. Simulated call data for Tuesday through Friday are provided. You may consider assigning different days of the week to different groups of students and having them report back to the class on their results.

➤ For projects, extensions, and more simulated data, go to (www.hsor.org).

TEACHER NOTES

SOLUTIONS

Part One: Using Data to Analyze a Business Problem

1. 17 shirts ($100 = 6x, x = 16.6667$)

2. 67 shirts ($300 + $100 = 6x, x = 66.6667$)

3. 52 shirts

4. $212 ($52 \cdot 6 - $100 = $312 - 100$)

5. No; their goal is to make $3 \cdot $100 = 300 profit. They made $212; therefore, they need to make at least $88 more to reach their profit goal.

6. 3.47 calls per hour (52 calls/15 hours of work = 3.47 calls per hour)

7. 349.1 minutes (59.4 + 75.6 + 72.1 + 65.2 + 76.8)

8. 6.7 minutes per order (349.1 total minutes taking orders/52 orders ≈ 6.7 minutes per order)

9. 6.3 minutes

10. 900 minutes (3 hours per day · 5 work days · 60 minutes per hour = 900 minutes)

11. 38.1% time busy ((349.1 − 6.3 minutes taking orders)/900 total work minutes = 0.380889)

12. When the phones are busy, any calls that get the busy signal will not be taken. Therefore, potential business is lost.

13. 38.1% (same as #11)

14. 61.9% of the calls (1 − 0.381 = 0.619)

15. 84 calls (52 calls = 0.619x, x = 84 calls)

16. 32 calls missed; (32) · ($6) = $192 in added marginal revenue

17.

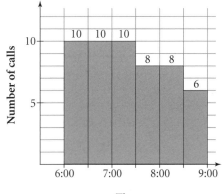

18. Decreasing over time; skewed right—the frequencies generally decrease as time progresses.

19. (a) is most likely, because the histogram shows that 30 calls were received before 7:30.0 and only 22 calls were received after 7:30.0.

20.

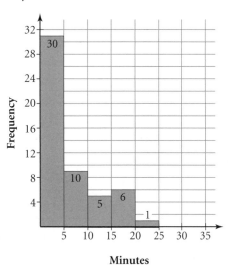

21. Skewed right—the frequency generally decreases as the length of call increases. The average duration of a call is 6.7 minutes (see the answer to question #8). See the histogram in the answer to question #20.

22. Choice (b) is most likely. It is clear from the graph that at least 30 calls were shorter than the average. In fact, an examination of the raw data shows that of the 52 calls placed during the sample week, 31 calls were shorter than 6.7 minutes, the average call length.

23. The median duration is 4.3 minutes. Notice that the 26th and 27th duration lengths are 4.3 minutes.

24. Choice (c) is most likely, because half of the calls have a duration shorter than 4.3 minutes and half have a longer duration.

25. The median is less than the mean in a skewed-right distribution.

TEACHER NOTES

Part Two: Analyzing Simulated Data to Manage Telephone Orders (see table below)

1. The phone line must not be in use.

2. The caller received a busy signal because a previous caller was being helped. No; there is no service time to be recorded for missed calls.

3. 8

4. 21.6 minutes; call #2 ended at 6:59.8, so call #3, which came in at 6:38.2, was placed on call waiting.

5. The call would be answered at 6:59.8 and would last 1.3 minutes.

6. Because the calls are made while a previous call is being served and another call is already waiting. To see that in the table, notice that call #2 has an end time later than the arrival

SIMULATED CALL DATA FOR MONDAY

Call #	Time of Call	Duration, Actual or Potential (in minutes)	One Line (no call waiting)		One Line (with call waiting)			Two Lines (no call waiting)		
			End Time	M	End Time	Wait Time	M	End Time 1	End Time 2	M
1	6:09.9	1.3	6:11.2		6:11.2			6:11.2		
2	6:36.8	23.0	6:59.8		6:59.8			6:59.8		
3	6:38.2	1.3		M	7:01.1	21.6			6:39.5	
4	6:42.2	6.6		M			M		6:48.8	
5	6:45.3	2.2		M			M			M
6	6:59.0	12.1		M			M		7:11.1	
7	6:59.4	15.0		M			M			M
8	7:07.0	2.2	7:09.2		7:09.2			7:09.2		
9	7:23.4	12.9	7:36.3		7:36.3			7:36.3		
10	7:27.3	14.4		M	7:50.7	9.0			7:41.7	
11	7:45.3	8.3	7:53.6		7:59.0	5.4		7:53.6		
12	7:49.0	1.9		M			M		7:50.9	
13	7:50.1	2.7		M			M			M
14	8:22.0	8.1	8:30.1		8:30.1			8:30.1		
15	8:30.9	1.8	8:32.7		8:32.7			8:32.7		
16	8:38.5	1.8	8:40.3		8:40.3			8:40.3		

time of each of these calls and call #3 has already been placed on call waiting. Therefore, calls #4 through #7 could not be answered.

7. Yes; 7:36.3

8. 9.0 minutes (7:36.3 − 7:27.3)

9. Call waiting; 7:59.0

10. Yes; two more calls could be taken due to the call waiting feature. Six calls would still be missed because another call was already on hold.

11. 3; average wait time is 12 minutes. Answers to the final question will vary as to whether customers would wait this long.

12. With two lines, call #3 would not have to wait. It would be picked up by line 2.

13. If both lines were busy, a call would still be missed.

14. 5

15. No call waiting: 8; call waiting: 10; two lines: 13

16. No call waiting: $48; call waiting: $60; two lines: $78

17. Based on these costs, recommend two phone lines, because that option increases Monday's estimated revenue by $30 with an added weekly cost of only $15. by comparison, call waiting increases Monday's estimated revenue by $12 with an added weekly cost of only $10.

However, workload concerns might override the profit advantage of two lines, since two workers would be needed every evening.

18. If we assume that Monday is typical of the average weekday, then the average weekly revenue is estimated to be 5($78) = $390. The weekly overhead is $100 (base) + $15 (second line). Their net is ($390 − $115) = $275. They miss their combined $300 target by an average of $25 per week. Also, having two lines requires that two of the partners work each night to staff the phone. They would certainly want to earn more money. This weekday time commitment might be unworkable without bringing in a fourth partner or paying someone to answer phones. This drawback might make call waiting the most attractive alternative.

19. Monday may be an atypical weeknight. In addition, the data represent a sample of just one Monday. (Data for Tuesday through Friday are provided, and could be analyzed as an extension.) In an actual study, a business would simulate every day of the week over a multiweek period. The business would also be concerned about seasonal variation in demand.

PROJECT: SIMULATION IN THE WORKPLACE

Students might want to write a letter to AT&T asking how their Call Processing Simulator works, which businesses have used it, and how many business simulations are run in an average year. Other suggestions include writing to the Federal Aviation Administration, an airport, or a major airline carrier (e.g., American, United, or Delta); a major telephone company; a state highway department or corrections department; a major department store or supermarket chain (e.g., Wal-Mart, K-Mart, or Kroger); or a branch of the military (e.g., U.S. Air Force, Army, Coast Guard, Marine Corps, or Navy).

Activity 9: Torn Shirts, Inc.

Simulated Call Data for Monday

Call #	Time of Call	Duration, Actual or Potential (in minutes)	One Line (no call waiting)		One Line (with call waiting)			Two Lines (no call waiting)		
			End Time	M	End Time	Wait Time	M	End Time 1	End Time 2	M
1	6:09.9	1.3	6:11.2		6:11.2			6:11.2		
2	6:36.8	23.0	6:59.8		6:59.8			6:59.8		
3	6:38.2	1.3		M	7:01.1	21.6			6:39.5	
4	6:42.2	6.6		M			M		6:48.8	
5	6:45.3	2.2		M			M			M
6	6:59.0	12.1		M			M		7:11.1	
7	6:59.4	15.0		M			M			M
8	7:07.0	2.2	7:09.2		7:09.2			7:09.2		
9	7:23.4	12.9								
10	7:27.3	14.4								
11	7:45.3	8.3								
12	7:49.0	1.9								
13	7:50.1	2.7								
14	8:22.0	8.1								
15	8:30.9	1.8								
16	8:38.5	1.8								

Does This Line Ever Move? © 2005 Key Curriculum Press

Activity 9: Torn Shirts, Inc.

Extension: Simulated Call Data for Tuesday

Call #	Time of Call	Duration, Actual or Potential (in minutes)	One Line (no call waiting)		One Line (with call waiting)			Two Lines (no call waiting)		
			End Time	M	End Time	Wait Time	M	End Time 1	End Time 2	M
1	6:00.3	4.1								
2	6:17.4	1.5								
3	6:36.6	4.7								
4	6:54.9	2.9								
5	6:56.8	7.0								
6	7:03.6	12.6								
7	7:04.0	17.8								
8	7:21.2	2.8								
9	7:26.6	4.0								
10	7:30.9	12.6								
11	7:33.1	4.1								
12	7:37.5	2.1								
13	7:42.3	5.6								
14	8:12.2	7.6								
15	8:28.7	19.6								
16	8:39.7	1.3								
17	8:56.2	3.1								
18	8:59.1	10.7								

Activity 9: Torn Shirts, Inc.

Extension: Simulated Call Data for Wednesday

Call #	Time of Call	Duration, Actual or Potential (in minutes)	One Line (no call waiting)		One Line (with call waiting)			Two Lines (no call waiting)		
			End Time	M	End Time	Wait Time	M	End Time 1	End Time 2	M
1	6:11.8	7.6								
2	6:17.4	6.2								
3	6:22.1	9.1								
4	6:31.3	3.5								
5	6:52.4	1.7								
6	6:57.7	16.2								
7	6:58.9	1.1								
8	7:14.2	1.5								
9	7:18.5	4.7								
10	7:18.6	7.9								
11	7:24.5	4.2								
12	7:29.7	11.0								
13	7:33.3	1.5								
14	7:40.3	14.2								
15	8:19.1	5.4								
16	8:30.3	4.3								
17	8:52.9	2.9								

Activity 9: Torn Shirts, Inc.

Extension: Simulated Call Data for Thursday

Call #	Time of Call	Duration, Actual or Potential (in minutes)	One Line (no call waiting)		One Line (with call waiting)			Two Lines (no call waiting)		
			End Time	M	End Time	Wait Time	M	End Time 1	End Time 2	M
1	6:05.1	1.7								
2	6:14.8	1.2								
3	6:22.0	18.1								
4	6:35.5	6.2								
5	6:38.1	7.2								
6	6:46.2	11.2								
7	6:51.8	2.0								
8	7:10.2	2.3								
9	7:33.4	4.3								
10	7:34.0	2.8								
11	7:38.3	4.0								
12	7:39.8	8.5								
13	7:41.6	3.7								
14	7:43.5	20.0								
15	7:46.4	1.3								
16	8:06.9	1.2								
17	8:08.0	15.4								
18	8:29.9	1.2								

Activity 9: Torn Shirts, Inc.

Extension: Simulated Call Data for Friday

Call #	Time of Call	Duration, Actual or Potential (in minutes)	One Line (no call waiting)		One Line (with call waiting)			Two Lines (no call waiting)		
			End Time	M	End Time	Wait Time	M	End Time 1	End Time 2	M
1	6:01.4	7.3								
2	6:06.2	6.0								
3	6:24.4	9.8								
4	6:26.3	1.5								
5	6:29.4	21.9								
6	6:31.2	3.4								
7	6:31.7	1.0								
8	6:39.7	3.6								
9	6:45.7	4.5								
10	6:53.9	15.7								
11	6:55.7	3.6								
12	7:09.1	1.8								
13	7:35.2	1.4								
14	7:48.3	2.3								
15	7:50.9	1.7								
16	7:52.1	12.4								
17	7:52.6	3.5								
18	8:09.1	7.8								
19	8:23.7	7.6								
20	8:26.7	1.8								
21	8:51.2	15.1								
22	8:57.6	1.0								

Decision Making Using Multiple Criteria

BACKGROUND

We all face decisions in our jobs, in our communities, and in our personal lives such as:

- Which college should I attend?

- Which job should I accept?

- Which car, house, computer, or stereo should I buy?

- Which health plan should I choose?

- Which supplier or building contractor should we hire?

- Where should a new airport, manufacturing plant, power plant, or health care clinic be located?

These decisions involve comparing alternatives that have strengths and weaknesses with regard to multiple objectives of interest to the decision maker. **Multiple-criteria decision making** is a structured methodology designed to handle the trade-offs inherent in making a decision that involves multiple criteria. One of the first applications of this methodology involved a study of alternative locations for a new airport in Mexico City in the early 1970s. The factors that were considered included cost, capacity, access time to the airport, safety, social disruption, and noise pollution.

This decision-making theory is a systematic approach for quantifying an individual's preferences. It is used to rescale a numeric value on some measure of interest onto a 0-to-1 scale, with 0 representing the least preferred option and 1 the most preferred option. Having a scale allows the direct comparison of many diverse measures. The next step is to assign a weight to each measure so that the rescaled measures can be combined into a single aggregate score. That is, with the right tools, it really is possible to compare apples to oranges! The end result is a rank-ordered evaluation of alternatives that reflects the decision maker's preferences.

An analogous situation arises when individuals, college sports teams, MBA degree programs, or even hospitals are ranked in terms of their performance on multiple disparate measures. For example, the Bowl Championship Series (BCS) in college football attempts to identify the two best college football teams in the United States to play in a national championship bowl game. Use of a multiple-criteria decision-making process has reduced (though not eliminated) the annual end-of-the-year arguments as to which college should be crowned the national champion.

CASE STUDY: STRATEGIC DECISIONS AT BC HYDRO

BC Hydro is a major supplier of hydroelectric power in British Columbia, Canada. The company faced many complex strategic decisions in the 1990s, including the addition of resources to generate electricity, the construction of transmission lines, and the negotiation of power agreements. The directors wanted to prepare to make those decisions in the best possible way, using quality information and sound logic in a coordinated manner.

The director of planning, realizing that all the decisions should contribute to achieving a set of long-range objectives, worked with operations research consultants, using multiple-criteria decision making, to identify the organization's strategic alternatives. Key executives were interviewed to determine their fundamental objectives and the trade-offs among objectives. The conclusion was that economics, environment, health and safety, and service quality were the major objectives that mattered.

References

Keeney, Ralph L., and Timothy L. McDaniels. 1992. Value focused thinking about strategic decisions at BC Hydro. *Interfaces* 22(6):94–109.

Choosing a College

THE SCENARIO

Miguel Ramirez, a high school senior, applied to several colleges and has been accepted at four: State University, which is far away; Podunk University, a small school where he is sure to receive a lot of academic support; and two schools with strong engineering reputations: I.Q.U. and Poly Tech Institute. Now he has to decide which offer to accept. Miguel asks his friend Chloe Aikens to help him think through this important decision. They both realize that there are many different issues to consider.

In this activity, you will see how Miguel and Chloe used a systematic process called **multiple-criteria decision making** to help Miguel make a thoughtful choice about the college he should attend. At the same time, you will apply this process to your own search for a preferred college.

Specify Criteria

With Chloe's help, Miguel has decided that academics, cost, location, and social life are the **criteria,** or factors, that are most critical in his choice of a college.

1. Generate a list of criteria that are important to you in choosing a college. You may choose as many criteria as you wish. When you have created your own list, compare lists with a partner. How are your lists similar? How are they different?

Next, Miguel and Chloe specified two or three measures for each criterion, as shown in the table.

Criterion	Measures
Academics	Average Combined SAT Score (of last year's freshman class)
	US News & World Report Rank
Cost	Living Expenses
	Tuition and Fees
Location	Average Daily High Temperature
	Nearness to Home
Social life	Athletics
	Reputation
	Size

Scale Each Measure

Next, Miguel and Chloe chose an appropriate scale for each of the nine measures. They also realized that some of the measures, such as average SAT score, have a **natural scale** (in this case, the combined score), while other measures, such as athletics, require a **constructed scale.** Furthermore, some of the measures are numeric (such as SAT scores), while others use categories. For example, Miguel and Chloe developed a scale of 1 to 4 for athletics:

1 Top 10 ranking in men's basketball or football or women's volleyball or basketball in the past two years

2 Top 25 ranking in any two of the sports above

3 Division I status

4 Other

They also realized that the range of each scale is important. For example, the theoretical range of the average combined SAT score is 400–1600, but realistically, Miguel and Chloe decided that a range of 900–1400 was sufficient. The scale range and type that they used for each measure are given in this table.

TYPES AND RANGES OF MEASURES

Measure	Range of Scale	Type
Average Combined SAT Score	900–1400 (realistic)	Numeric-natural
US News & World Report **Rank**	1. Top 25	Categorical-constructed
	2. 26–100	
	3. 101–250	
	4. Unranked	
Living Expenses	$6,000–$12,000 (realistic)	Numeric-natural
Tuition and Fees	$5,000–$20,000 (realistic)	Numeric-natural
Avg. Daily High Temperature	50°–70°F	Numeric-natural
Nearness to Home	1. Commuting (0–100 miles)	Categorical-constructed
	2. Within 4-hr. drive (101–250 miles)	
	3. Within day's drive (251–500 miles)	
	4. Far (over 500 miles)	
Athletics	1. Top 10 in one sport	Categorical-constructed
	2. Top 25 in two sports	
	3. Division I	
	4. Other	

(table continued on next page)

TYPES AND RANGES OF MEASURES (continued)

Measure	Range of Scale	Type
Reputation	1. Seriously academic	Categorical-constructed
	2. Balanced academics and social life	
	3. Party school	
Size	1. Under 3,000	Categorical-constructed
	2. 3,001–6,000	
	3. 6,001–12,000	
	4. Over 12,000	

2. Define a measure for each of your own criteria. Try to limit yourself to no more than three measures per criterion. To ensure a manageable number of choices, it is best to narrow your selection of colleges to no more than five. Include at least two colleges that you are fairly certain you will be accepted to. Your list should represent the diversity of your preferences. For example, it may include a small college and a large university, or a nearby school and one far from home. Create a table showing a scale or range for each of the measures you listed. Remember that you will need to construct a scale for any measure that does not have a natural scale. Label each scale as natural or constructed.

Collect Data

After scaling each measure, Miguel and Chloe collected the data shown in this table.

MEASURES FOR MIGUEL'S FOUR COLLEGES

Measure	Podunk U.	State U.	I.Q.U.	Poly Tech
Average Combined SAT Score	1050	1000	1300	1320
US News & World Report **Rank**	Unranked	200	30	20
Living Expenses	$9,000	$7,000	$9,000	$12,000
Tuition and Fees	$18,000	$6,000	$8,000	$12,000
Average Daily High Temperature	55°F	58°F	62°F	68°F
Nearness to Home	4	1	3	4
Athletics	4	1	3	3
Reputation	2	3	1	1
Size	1	4	3	2

3. Collect data and construct a similar table for your choice of colleges. Sources for these data can be the Internet, a commercial college selection text, or your school's college counselors.

Rescale to Common Units

Once Miguel and Chloe collected the data, Chloe reminded Miguel that if they compared the data in its current form, it would be like comparing apples to oranges. They decided to convert the data to "common units." This means deciding what the most preferred value of each measure is, assigning a value of 1 to it, deciding what the corresponding least preferred value of each measure is and assigning a value of 0 to it. For intermediate values, if the measure has a numeric scale, the common unit value may be assigned proportionally. For example, the average combined SAT score at Podunk U. was 1050. The range for this measure was 900–1400, so 900 converted to 0, 1400 to 1, and the proportional value for Podunk U. was $(1050 - 900)/(1400 - 900) = 0.3$.

On the other hand, for categorical measures, after assigning the best value a common unit of 1 and the least value a common unit of 0, Miguel and Chloe had to decide how to allot the common units. In some cases, allotment might be proportional, but in other cases it might not be. They decided to use proportional common units for every categorical measure.

Some of the results of Miguel's and Chloe's rescaling to common units appear in the Conversion to Common Units table. Refer to the Types and Ranges of Measures table and fill in the rescaled common units for each of the blanks in the table below.

CONVERSION TO COMMON UNITS

Measure	Podunk U.	State U.	I.Q.U.	Poly Tech
Average Combined SAT Score	0.3			
US News & World Report **Rank**	0.0	0.33	0.67	1.0
Living Expenses	0.5			0.0
Tuition and Fees		0.93		
Average Daily High Temperature	0.25			
Nearness to Home			0.67	
Athletics			0.33	0.33
Reputation	1.0	0.0		
Size	0.0	1.0		

4. Complete a similar table of conversion to common units for the real data you collected for your own choice of colleges.

Conduct an Interview to Calculate Weights

Next, in order to assign a weight to each of the measures to reflect the relative importance Miguel attaches to each of them, they decided that Chloe should interview Miguel. She made observations to ensure that Miguel understood the measures and the effects of the assigned weights. They then created this table.

RANKING AND WEIGHTING MEASURES

Criterion	Measure	Least Preferred	Most Preferred	Rank Order	Points (0–100)	Calculated Weight (points/sum)
Academics	**Average Combined SAT Score**	900	1400			
	US News & World Report **Rank**	Unranked	Top 25			
Cost	**Living Expenses**	$12,000	$6,000			
	Tuition and Fees	$20,000	$5,000			
Location	**Avg. Daily High Temperature**	50°F	70°F			
	Nearness to Home	Far	Very close			
Social Life	**Athletics**	Other	Top 10			
	Reputation	Party	Balanced			
	Size	Under 3,000	Over 12,000			
					Sum	

5. Using the data for your personal college choice, create a table similar to the Ranking and Weighting Measures table. Be sure to include your own most preferred and least preferred values for each of your measures.

Chloe: We have some measures and their ranges for making a decision about your college preference. Focus first on the column of least preferred values. For which one of the measures is it most important to you to increase from the least preferred value to its most preferred value? For example, is it more important to you to increase the SAT score from 900 to 1400 or to reduce tuition from $20,000 to $5,000?

Miguel: Lower the tuition!

Chloe: Are you sure that lowering the tuition to $5,000 is the most important improvement in the whole list?

Miguel: Yes. I think we should rank tuition number 1.

Chloe: What would be the next most important measure to move from least to most preferred?

Miguel: *US News & World Report* rank is important. So let's rank that second, and SAT score third.

Chloe: The next goal is to subjectively assign points from 0 to 100 for each measure based upon the rank order, where rank order is the relative importance of each measure to you. Start by assigning 100 points to the tuition range, which you ranked first. Now, you ranked the *US News & World Report* rank second. How important is this rating from least to most preferred, compared to reducing the cost of tuition from $20,000 to $5,000? If it's close, you should use a number close to 100.

Miguel: I think it's about 90% as important, so let's use 90 points for that one, and SAT scores are almost as important, so we'll use 88 points for that range.

Miguel and Chloe continued this process until they had ranked all nine measures and assigned a relative score to each. Notice that reputation and nearness to home are both ranked sixth because Miguel feels they are equally important.

6. Just as Chloe interviewed Miguel, exchange interview roles with a partner to rank your own measures and assign points to them. The next table contains the points Miguel assigned to each of his measures. Assign points to your own measures based on the rank order you placed them in and the relative importance to you of having a value closer to the most preferred value over the least preferred value.

MIGUEL'S RANK AND POINT ASSIGNMENT

Criterion	Measure	Least Preferred	Most Preferred	Rank Order	Points (0–100)	Calculated Weight (points/sum)
Academics	**Average Combined SAT Score**	900	1400	3	88	
	US News & World Report **Rank**	Unranked	Top 25	2	90	
Cost	**Living Expenses**	$12,000	$6,000	4	70	
	Tuition and Fees	$20,000	$5,000	1	100	
Location	**Avg. Daily High Temperature**	50°F	70°F	8	20	
	Nearness to Home	Far	Very close	6	50	
Social Life	**Athletics**	Other	Top 10	5	60	
	Reputation	Party	Balanced	6	50	
	Size	Under 3,000	Over 12,000	9	10	
				Sum	538	

Chloe: Miguel, what did you get for the total number of points for all your measures? Once you have the point total, you'll need to divide the points for each measure by this total to get the weight.

Miguel: I got 538 total points. Now I am going to calculate the weight of each measure and enter the value in the last column of my Rank and Point Assignment table.

Chloe: Miguel, what is the total weight for each set of criteria?

Miguel: When I combine the weights for the measures of each criterion, I get a total of 0.33 for academics, 0.32 for cost, 0.13 for location, and 0.22 for social life.

Chloe: Which set has the greatest weight assigned to it?

Miguel: It looks like academics, with 0.33.

Chloe: Are there criteria with similar weights?

Miguel: It looks like academics and cost are almost the same.

Chloe: Are these the criteria you feel are the most important to you in choosing a college, and do you think they're about the same in importance?

Miguel: I didn't realize I placed so much importance on academics.

Chloe: What did you expect to happen?

Miguel: I thought the social life would be at the top of the list!

Chloe: Well, you gave athletics 60 points, reputation 50 points, and size only 10 points. Do you want to change anything?

Miguel: No, I really think academics and cost are most important.

7. Calculate to two decimal places the weight of Miguel's measures and record them in the last column of Miguel's Rank and Point Assignment table.

   Using the same process, compute the weights for each of your own measures. Check to see if the sum of the weights is close to 1.

8. Discuss these questions with your classmates:

   - Which of your criteria has the greatest weight?
   - Is this the criterion that you feel is the most important for choosing a college? If not, explain why your results are different from what you expected.
   - Is the number of measures an indication of a criterion's importance? Explain.
   - Did the criterion with the most measures receive the most points? Explain whether you agree or disagree that this particular criterion is most important to you in choosing a college.

- Are there any measures or criteria you omitted that you now think should have been included? If so, revise your list and start over again. The process should move quickly, since many of your responses will be close to what you answered before.

Calculate Total Scores

Next, Miguel and Chloe determined the college that is his preferred choice. They used the data from the Conversion to Common Units table, in which common units are computed, and the weights calculated in the last column of Miguel's Rank and Point Assignment table. To calculate the total scores, Miguel calculated the product of the weight (W) and the corresponding common unit (CU): Score $= W \times CU$ (rounded to three decimal places). By totaling the points for each college, Miguel learned which of his college choices best suits his needs. His results are shown in the next table.

TOTAL SCORES OF MIGUEL'S COLLEGES

Measure	Weight	Podunk U.	State U.	I.Q.U.	Poly Tech
Avg. Combined SAT	.16	.16 × .3 = .048	.032	.128	.134
US News Rank	.17	.000	.056	.114	.170
Living Expenses	.13	.065	.108	.065	.000
Tuition and Fees	.19	.025	.177	.152	.101
Avg. Daily High Temp.	.04	.010	.016	.024	.036
Nearness to Home	.09	.090	.000	.060	.090
Athletics	.11	.000	.110	.036	.036
Reputation	.09	.090	.000	.060	.030
Size	.02	.000	.020	.010	.010
TOTAL SCORE	1.00	.328	.519	.649	.607

9. Miguel reviewed the information in the table and found that I.Q.U. and Poly Tech were ranked 1 and 2 by his weighting preferences. He decided to look more closely at the numbers and compare the strengths and weaknesses of the two colleges. On which measures was I.Q.U. significantly better than Poly Tech? Significantly worse? About the same? Based on this table and his subsequent review of the strengths and weaknesses, Miguel has decided to attend I.Q.U.

10. Use this same procedure with your data to convert your common units and weights to a score for each measure. Then obtain a total score for each of your colleges. Based on your scores, which college should you choose?

EXTENSION: THE RIGHT CAR FOR YOU

You are about to purchase a used car and have decided to apply multiple-criteria decision making to choose which car to buy.

1. Define the primary objective of your process.

2. Identify the criteria that will contribute to your selection.

3. For each criterion, identify one or more measures.

4. For each measure, create a scale. Be careful to use a realistic range.

   Suggestions: Regarding the soundness of the motor, a categorical scale could be

   - in excellent condition
   - needs minor repairs
   - burns oil, but still runs well
   - starts

5. Collect data on each of the vehicles you are considering.

   It may be appropriate to approach the seller of the vehicle with a list of questions that cover the measures you have selected.

6. Convert the data on each measure to common units, similar to the Conversion to Common Units table in the activity.

7. Rank, assign points, and determine the weight of each measure, as in the Ranking and Weighting Measures tables in the activity.

8. Calculate a total score for each vehicle, as in the Total Scores of Miguel's Colleges table.

PROJECT: DECISION MAKING IN LOCAL AND STATE GOVERNMENT

Identify a local or state government decision that involves multiple objectives, such as the location of a government facility or the prioritization of projects (e.g., which highway to repair or which hazardous waste site to clean up first). Research the decision-making process and the relevant background. Then apply what you know about decision making using multiple criteria.

- Describe how the decision was made and the most important issues that drove the final choice.

- Identify a list of objectives that could have been used to make the decision.

- Create measures for each of the objectives and realistic ranges.

- Rank the order and weight the measures according to a team consensus.

- How might different groups of people (that is, special-interest groups) have different objectives and weights?

10

Choosing a College

The broadest objective of this lesson is to demonstrate for students a process that brings the power of mathematical reasoning to bear on a difficult decision involving multiple criteria. Perhaps the most interesting feature of this process is that there is no one correct answer. The preferred alternative will depend on individual preferences that are reflected in the variables and measures of those variables that are chosen, as well as in the weights that are assigned to each of those measures. Yet the process brings a certain amount of objectivity to decision making.

OBJECTIVES

- identify key variables, and create appropriate measures and scales
- distinguish between categorical and numeric data, natural and constructed scales
- experience multiple-criteria decision making in a relevant context

TECHNOLOGY NOTE

Spreadsheets or graphing calculators with lists can facilitate the calculations.

INITIATING THE ACTIVITY

Ask students to bring data on their favorite colleges to class for this lesson (college brochures, catalogs, information from websites).

To help students begin to think about the issues that are important in the activity, we strongly recommend using these discussion questions to introduce and clarify the decision-making process.

1. How would you go about choosing a college? (The idea of multiple criteria should emerge from this discussion.)

2. Discuss problems you might encounter in trying to accommodate all of the criteria. (The need for a formal process should emerge here.)

3. Create a list, posting items as they are suggested, of criteria you might need to consider when choosing a college. (Encourage students to identify at least one criterion, such as social life, which is not measurable in an obvious way.)

4. As a class, reach a consensus on only six to eight criteria from the list to use, based on what students consider important. Decide whether any of the criteria can be grouped in a category. Reach a class consensus on the categories.

5. Explore how to measure two of the criteria.

6. Discuss how to create a measure for a criterion such as social life. (This would likely be some sort of categorical scale. For example, a four-point categorical scale for social life could be (1) poor, (2) fair, (3) good, and (4) very good.)

7. For the two measures selected earlier, determine a realistic range relevant to your choices. (For example, combined SAT scores range between 400 and 1600. However, a student might consider only those colleges where the average score for an entering class ranges from 900 to 1300. This would be a more realistic range for this measure.)

8. Try to compare the scales for the two measures. Does this create any problems? (Because one would likely be comparing apples to oranges, this question should set up the need for some sort of common scale.)

9. Ask each student to rank the criteria in order of importance from most to least important. In groups of three, have students share their ranked lists and discuss how they vary. What does this variation indicate? Explore why different students would end up with different rank orderings of a list of colleges.

It is important to emphasize that the "right" school will vary from person to person. As a final wrap-up, ask students to identify other decisions they might face that involve multiple criteria. Possibilities include buying a car and accepting a part-time job.

GUIDING THE ACTIVITY

This activity gives students a method for making informed decisions involving many criteria. The scenario models a process for the students to follow as they choose their own colleges. You might have students complete the activity over the course of two or three days.

Remind students that scaling is arbitrary. Both constructed and natural scales are completely subjective. For example, one student may desire to be closer to home, whereas another may prefer to be farther away. Individual biases will influence both scales and their ranges.

The discussion of a realistic range for SAT scores in the scenario is a critical issue. Failure to use an appropriate range can affect the results. For example, if we work with the natural range for SAT scores, 400 to 1600, State U.'s raw score of 1000 converts to a common unit score of 0.5, compared to a common unit score of 0.75 for I.Q.U.'s raw score of 1300. In that case, the common unit score for I.Q.U. is only 50% greater than the common unit score for State U. If we instead use a more realistic range of 900 to 1400, the corresponding common unit scores are 0.2 and 0.8, a more dramatic difference.

TEACHER NOTES

Multiple-criteria decision-making terminology identifies measures as continuous or categorical and as natural or constructed. However, many times measures that would be identified as continuous are not really continuous in the sense of being on a continuum of real numbers. In order to avoid possible student misunderstandings, we have used the term "numerical" in place of "continuous."

The other type of measure is called "categorical-constructed." Because this type of measure does not have a natural scale, a set of categories must be constructed. For example, Miguel listed nearness to home as a measure and chose to break it into four categories, each with a common unit value from 0 to 1. The assignment of the point values from 0 to 1 is completely subjective. Miguel did not want to attend a college close to home, so he assigned schools in category 1 (within 100 miles) a common unit of 0. In all of the categorical measures, Miguel chose to scale the categories proportionally. Students should be prepared to justify their point assignments. The completed table of Conversion to Common Units is shown here.

CONVERSION TO COMMON UNITS

Measure	Podunk U.	State U.	I.Q.U.	Poly Tech
Average Combined SAT Score	.3	.2	.8	.84
US News & World Report **Rank**	0	.33	.67	1
Living Expenses	.5	.83	.50	0
Tuition and Fees	13	.93	.80	.53
Average Daily High Temperature	.25	.4	.6	.9
Nearness to Home	1	0	.67	1
Athletics	0	1	.33	.33
Reputation	1	0	.5	.5
Size	0	1	.67	.33

The calculated weights for Miguel's criteria are, from top to bottom, .16, .17, .13, .19, .04, .09, .11, .09, and .02. These weights are then used in the table Total Scores of Miguel's Colleges. The weights may total slightly more or less than 1 due to round-off error.

You may want each student to write an analysis of his or her findings, explaining why the college that was ranked number 1 was at the top of his or her list. Students might also discuss external factors that would force a person to choose a college other than the top-ranked college.

➤ For more teacher notes and extensions, go to (**www.hsor.org**).

Activity 10: Choosing a College

Types and Ranges of Measures

Measure	Range of Scale	Type
Average Combined SAT Score	900–1400 (realistic)	Numeric-natural
US News & World Report **Rank**	1. Top 25 2. 26–100 3. 101–250 4. Unranked	Categorical-constructed
Living Expenses	$6,000–$12,000 (realistic)	Numeric-natural
Tuition and Fees	$5,000–$20,000 (realistic)	Numeric-natural
Avg. Daily High Temperature	50°–70°F	Numeric-natural
Nearness to Home	1. Commuting (0–100 miles) 2. Within 4-hr. drive (101–250 miles) 3. Within day's drive (251–500 miles) 4. Far (over 500 miles)	Categorical-constructed
Athletics	1. Top 10 in one sport 2. Top 25 in two sports 3. Division I 4. Other	Categorical-constructed
Reputation	1. Seriously academic 2. Balanced academics and social life 3. Party school	Categorical-constructed
Size	1. Under 3,000 2. 3,001–6,000 3. 6,001–12,000 4. Over 12,000	Categorical-constructed

Activity 10: Choosing a College

Measures for Miguel's Four Colleges

Measure	Podunk U.	State U.	I.Q.U.	Poly Tech
Average Combined SAT Score	1050	1000	1300	1320
US News & World Report Rank	Unranked	200	30	20
Living Expenses	$9,000	$7,000	$9,000	$12,000
Tuition and Fees	$18,000	$6,000	$8,000	$12,000
Average Daily High Temperature	55°F	58°F	62°F	68°F
Nearness to Home	4	1	3	4
Athletics	4	1	3	3
Reputation	2	3	1	1
Size	1	4	3	2

Activity 10: Choosing a College

Conversion to Common Units

Measure	Podunk U.	State U.	I.Q.U.	Poly Tech
Average Combined SAT Score	0.3			
US News & World Report Rank	0.0	0.33	0.67	1.0
Living Expenses	0.5			0.0
Tuition and Fees		0.93		
Average Daily High Temperature	0.25			
Nearness to Home			0.67	
Athletics			0.33	0.33
Reputation	1.0	0.0		
Size	0.0	1.0		

Activity 10: Choosing a College

Ranking and Weighting Measures

Criterion	Measure	Least Preferred	Most Preferred	Rank Order	Points (0–100)	Calculated Weight (points/sum)
Academics	Average Combined SAT Score	900	1400			
	US News & World Report Rank	Unranked	Top 25			
Cost	Living Expenses	$12,000	$6,000			
	Tuition and Fees	$20,000	$5,000			
Location	Avg. Daily High Temperature	50°F	70°F			
	Nearness to Home	Far	Very close			
Social Life	Athletics	Other	Top 10			
	Reputation	Party	Balanced			
	Size	Under 3,000	Over 12,000			
				Sum		

Activity 10: Choosing a College

Miguel's Rank and Point Assignment

Criterion	Measure	Least Preferred	Most Preferred	Rank Order	Points (0–100)	Calculated Weight (points/sum)
Academics	Average Combined SAT Score	900	1400	3	88	
	US News & World Report Rank	Unranked	Top 25	2	90	
Cost	Living Expenses	$12,000	$6,000	4	70	
	Tuition and Fees	$20,000	$5,000	1	100	
Location	Avg. Daily High Temperature	50°F	70°F	8	20	
	Nearness to Home	Far	Very close	6	50	
Social Life	Athletics	Other	Top 10	5	60	
	Reputation	Party	Balanced	6	50	
	Size	Under 3,000	Over 12,000	9	10	
				Sum	538	

Activity 10: Choosing a College

Total Scores of Miguel's Colleges

Measure	Weight	Podunk U.	State U.	I.Q.U.	Poly Tech
Avg. Combined SAT	.16	$.16 \times .3 = .048$	.032	.128	.134
US News Rank	.17	.000	.056	.114	.170
Living Expenses	.13	.065	.108	.065	.000
Tuition and Fees	.19	.025	.177	.152	.101
Avg. Daily High Temp.	.04	.010	.016	.024	.036
Nearness to Home	.09	.090	.000	.060	.090
Athletics	.11	.000	.110	.036	.036
Reputation	.09	.090	.000	.060	.030
Size	.02	.000	.020	.010	.010
TOTAL SCORE	1.00	.328	.519	.649	.607

Does This Line Ever Move? © 2005 Key Curriculum Press